# THIS ERRATIC PLANET

## WHAT HAPPENS WHEN THE EARTH CHANGES ITS AXIS OF ROTATION

Sir Ian Niall Rankin M.A. (Oxon)

© 2019 Ian Niall Rankin. All rights reserved.

No part of this book may be reproduced, stored in a retrieval system, or transmitted by any means without the written permission of the author.

Any people depicted in stock imagery provided by Thinkstock are models, and such images are being used for illustrative purposes only.
Certain stock imagery © Thinkstock.

This book is printed on acid-free paper.

Because of the dynamic nature of the Internet, any web addresses or links contained in this book may have changed since publication and may no longer be valid. The views expressed in this work are solely those of the author and do not necessarily reflect the views of the publisher, and the publisher hereby disclaims any responsibility for them.

*To my three children:*
*ZARA, GAVIN and LACHLAN*

# Contents

| | |
|---|---|
| **Preface** | **xi** |
| **PART I** | **1** |
| **The Earth is Going to Fall Over** | **1** |
| Background | 1 |
| **Uniformity Versus Catastrophe** | **1** |
| Evidence does not support gradualism | 1 |
| Sudden death of the mammoths | 3 |
| Other disasters have struck | 7 |
| Analysis of these deaths | 12 |
| Land disrupted | 13 |
| Volcanoes and earthquakes | 16 |
| Changing pattern of catastrophe | 19 |
| **What Caused this Earlier Violence?** | **21** |
| The Earth's unstable axis | 21 |
| The Sahara Desert | 25 |
| **Pole Shift** | **29** |
| The Earth's axis has many choices | 29 |
| Another choice of pole movement | 34 |
| **There Never Was an Ice Age** | **37** |
| The polar ice sheets simply moved with their poles | 37 |
| Earlier thoughts on ice ages | 41 |
| Ice-age theory doesn't fit the evidence | 46 |
| Sea levels | 48 |
| Sea levels with pole shift | 51 |
| Where were the last poles? | 52 |
| **The Earth's Magnetic Field** | **55** |
| Palaeomagnetics | 55 |
| What produces the Earth's magnetic field? | 58 |
| Relations between geographical and magnetic poles | 61 |

| | |
|---|---|
| The reversing magnetic field | 62 |
| Seafloor spreading | 64 |
| Recent magnetic reversals | 66 |
| **The Climate has Changed** | **69** |
| What is under the seafloor? | 69 |
| The Holocene weather | 74 |
| **History Speaks of Catastrophe** | **76** |
| Man writes of change | 76 |
| Some 3,500 years ago | 80 |
| 1000 BC to 500 BC | 85 |
| Earlier calendars | 87 |
| The 360-day year | 88 |
| **Have we Made a Case for Catastrophe?** | **91** |
| The distant past clearly indicates calamities | 91 |
| The dinosaurs | 92 |
| The doomsday process | 95 |
| Events surrounding a pole shift | 97 |
| What is the trigger mechanism of a pole shift? | 99 |
| The Earth cleanses itself | 101 |
| History is lost | 103 |
| Summary | 103 |
| Afterthought | 104 |
| People prior to the end of the 'ice age' | 104 |
| Megalithic evidence | 107 |
| Tiahuanaco again | 111 |

**PART II**      **114**
**There is Something Missing From the Solar System 114**

| | |
|---|---|
| Why won't the planets run out of energy? | 114 |
| What is gravity? | 119 |
| The New Theory | 122 |
| Why can't we see the Magnetic Centre? | 124 |

Sunspots                                        129
How the New Theory works                        131
Planetary locations                             134
The plane of the ecliptic                       136
The planets' satellites                         138
Comets                                          143
Spacecraft *Pioneer 10* and *11*                **146**
Asteroids                                       147
Summary                                         148
Conclusion                                      150
Evaluating the New Theory                       153
The solar system gives out a new feeling        154

**Appendix**                                    **155**
The origin of the solar system                  155

**Bibliography**                                **157**

**Index**                                       **170**

## *Preface*

This is a strange book. It is divided into two parts, which at first glance do not belong within the same cover. But another look shows that, without the second part, the larger first section would lose much of the supporting argument it needs to explain the violent, yet fragile, behaviour of the Earth.

This fragility is what the first part of the book is about. In it I hope to show that one of the Sun's satellites, in this case the Earth, is repeatedly changing its axis of rotation: it shifts its angle of tilt with respect to the ecliptic – the average plane of the Earth's orbit round the Sun – and it also changes the location of the North and South Poles. Needless to say, with those changes – often very rapid by human and not geological timescales – goes devastation. In the past, floods and storms have exceeded anything we know today; and low-level wind speeds have breached the 1999-recorded 350 mph.

Received thinking on this violence does not agree. In the early nineteenth century a change of thinking had come about whereby belief that the Earth's behaviour was sudden and very violent – a feature of the previous century – was replaced with a belief in gradualism: all major changes in the Earth's behaviour were slow and rarely abrupt. This of course ushered in Darwinism. I believe there is considerable evidence for change that is extreme and oft-repeated, not simply those changes that appear noticeable at the boundaries of the geological periods, such as the Jurassic or Cretaceous. This argument is supported by the discovery of animals which have died in a violent and unexpected way; by observation of land changes that have little gradualism about them; and by records of vulcanism.

What is indicated by sudden geological change is, of course, a change in climate; this in its turn leads to changes in the geological strata that are laid down. Desertification can be followed by what others choose to call an ice age. This book tries to show that what is deemed an ice age is something very different. Cold periods there have certainly been, but periods when the temperature of the Earth, worldwide, drops by many degrees are not demonstrable. I have tried to support this thinking with a close study of the Earth's magnetic field and its historical imprints supplied by palaeomagnetic readings. Subsequently, I look at some of the strange utterances of written history; the further back these go the less people are prepared to trust them, and the same descriptions of events are deemed fables.

Why, you may ask, are these catastrophic events not better known? The answer lies in the magnitude of these world changes. The Earth gets weary of staying in one position. Like a sleeping giant it has to turn over. Certain types of axis shift have meant destruction of human life on a lemming scale. With few humans left, technological development is lost without trace, and man has to revert to stone-age conditions. I wonder how often in the distant past human progress has achieved the technological levels of today, only to lose them for the next 300 generations.

The answer may be 'quite often'. The level of devastation I refer to is so extreme that archaeological remains are wiped out: it only takes a few metres of topsoil, moved over a site, for that relic of human enterprise to be concealed indefinitely. Our technological 'civilisation' has developed, since Neolithic times, over some 7,000 years. That is such a tiny snatch from the span of *Homo sapiens sapiens*. Surely we cannot be so arrogant as to think that this is the first time civilisation has reached its present levels. After all, we know that man's brain structure and capacity has been in its present form for several hundred thousand years, even if some contemporaries have had lesser cerebral powers. I hope the pointers in this book will suggest that man might well have been watching television in the last interglacial.

In the second part, I cannot claim as mine the curious push-pull New Theory that accounts for the movements of planets and other bodies within the solar system. I heard it from a South American, Osvaldo Pedrosa, who in his turn would claim that this thinking came from someone whose name he never discovered. Despite its questioned origin, the theory holds great charm, and seems to produce an explanation for happenings that previously had none. I can lay claim to much of the derived thinking; I have made adjustments; and I have added further thought on satellite behaviour. But the real 'invention' lies in the different relationship that the Sun has with its followers.

That this new approach has a startling impact on the role given to the Sun by received thinking will worry many conventional scientists, in spite of the fact that it explains much of what was contradictory, and much that still needed an explanation. That said, Newton's all-important inverse square law of force still holds good, up to a point; but the attraction between heavenly bodies is explained in a different way.

The New Theory introduces a revised attitude to the solar system. It is no longer a system of unstoppable, fast-flying satellites which were launched 4.6 billion or more years ago from the proto-Sun, and which gained their momentum at the time they were born. Instead, it becomes a system which is self-sustaining, where the satellites constantly draw their energy for movement from the centre of the solar system. Because they do not have the innate momentum previously attributed to them, they are far more impermanent, far more capable of changing.

In Part I, I make constant reference to the New Theory revealed in Part II. The unstable behaviour of the Earth needs an explanation that does not coincide with present beliefs in how the Earth operates. Then, too, the question of the Earth's rotation – unexplained by conventional thinking – needs answers. So it may appeal to the reader to jump forward to the second part, and see how important the New Theory is for opening up these unknowns.

# PART I

## *The Earth is Going to Fall Over*

**Background**

This book is about the impending disaster that is fast heading towards the Earth. Some scientists, because of the way the Earth's magnetic field is behaving, believe that some thirty years from now that magnetic field will no longer exist. We believe this signals disaster. Received science has always maintained that the angular momentum of the Earth rotating at the astonishingly slow speed of one revolution every twenty-four hours was such that it was impossible for the Earth to shift from its present axis of rotation. We believe that the New Theory, explained in Part II, changes that: although there is innate gyroscopic momentum in the Earth's rotation, it is quite possible for the forces causing rotation, under given circumstances, to affect the Earth in destabilising ways. This we will discuss. Let us start by going back to the thinking of the earliest geologists – in the nineteenth century.

## *Uniformity Versus Catastrophe*

**Evidence does not support gradualism**

The doctrine of uniformitarianism was first advanced by James Hutton in 1795; it took the view that the steady-state world we see from day to day had always been that way.

All changes had occurred gently and gradually. The earthquakes or volcanic explosions may have been many in the past but they were never dire. The explosion that blew apart the island of Santorini in the Aegean around 1500 BC was as large as anything that has ever occurred; that this had been responsible for bringing to a close the Minoan civilisation on Crete, some 110 km away, did not undermine uniformity theory. The doctrine came into its own when supported by Sir Charles Lyell (1797–1875) who was one of the most influential geologists of the early to middle nineteenth century; and it was on the back of this gradualism that Darwin, who was a close disciple of Lyell, hung *The Origin of Species*.

The theory of uniformity maintains that no processes took place in the past that are not taking place now. Mountains were pushed up very slowly, then wind, rain and ice slowly eroded them, carrying the detritus by river or glacier down to the sea, and turning the mountains into plains. Equally, at sea, wave action would bring down cliffs, and gradually move the coastline back, to be followed by a slowly rising seabed which would move the coastline out again. Volcanoes, too, though seemingly violent, were local incidents: dormant mountains would wake up, spew out lava and go to sleep again for years, only to begin the cycle again.

This gradualism does not fit the facts. There is strong evidence that within the last twenty thousand years the Earth has been subjected to violence that far exceeds anything permissible under the heading of uniformity. To provide evidence of abrupt change, we are going to look at events that occurred within a very short geological time span: the last ten thousand years of the Pleistocene epoch, and the ten thousand years making up the Holocene epoch, leading to the present. The events will include sudden climatic change, violent deaths of fauna, and abrupt movement in the Earth's crust. Only by reducing the time allowed for the duration of these events do we make a case for the type of extreme catastrophe which we believe has struck the Earth many many times – and several of those within the last twenty thousand years.

**Sudden death of the mammoths**

Let's start with the much-debated woolly mammoths of northern Siberia. The coastline of Siberia from Novaya Zemlya to its eastern extremity, and including the New Siberian Islands, Stolbovoi and Bel'kovskiy, was until recently littered with bones of mammoths; they did not all die at the same time – this has been established by Carbon-14 dating. However, they died in a terrain that is now uncultivable, and capable of growing little more than moss and the other dwarf plants that compose the tundra. What is certain is that the region would not be capable of supporting these animals nowadays, even in smaller numbers. So what has changed the climate? Then, too, there are other mammoths that have been found with flesh intact, so much so that the dogs pulling local sledges have eaten their flesh. The carcasses have been buried in the permafrost so rapidly that, in some instances, even the eyes have not decomposed; in other cases food has been found in the beasts' mouths. This, together with food from their stomachs, on analysis has turned out to be plants, grasses and leaves off trees that are now found in southern Siberia.

Charles Darwin, who would not admit the occurrence of catastrophes, in a letter to Sir Henry Howarth confessed that the extinction of mammoths and other mammals including rhinoceroses was for him an unexplained mystery. Disastrous freezing had struck, and what was doubly curious was that it appeared to have happened so fast that the animals had not had time to decompose. Carbon-14 dating has indicated that some of these animals suffered death and deep freezing as long ago as 50,000 years. There are others that have died under the same conditions even more recently than the end of the alleged last ice age. So it can be surmised that this type of sudden death has been inflicted on these northern regions with some degree of repetitiveness in the last part of the Pleistocene and Recent epochs.

Scientists concerned with cryogenics have suggested that the temperatures required to freeze living creatures with the speed described would be in the region of minus 150°F. That would be a very low reading, in excess of that found at either pole during the summer – it could however be achieved during a polar winter with a high wind chill factor.

Present-day examples of sudden freezing where animals have fallen into ice pockets have not led to flesh preservation such as is found in the Siberian permafrost. Sir Fred Hoyle in his book *Ice* says:

I have been informed that today, when reindeer fall down crevasses in the Greenland ice, they are subsequently found to be in an unpleasantly putrefied condition. It seems that, no matter how cold the air is, the body heat of the animal is sufficient to promote bacterial decomposition. The Siberian mammoths, in spite of their much greater body weight have not putrefied in the same way, which would support the suggestion that they were robbed of their body heat at an extremely rapid rate – much quicker than conduction into cold air could give.

Examples of nearly perfectly preserved specimens have been dated to 50,000 years BP (before the present), 44,000 BP, 20,000 BP, 11,500 BP, and 3,700 BP.[1] A recent discovery was in 1999: a woolly mammoth twenty thousand years old, given the name Zharkor, was found near the town of Khatanga, well north of the Siberian Arctic Circle, and is in an astonishing state of preservation. It will have much to tell when it is released from the block of ice that has been its home for so many millennia. It must be remembered that these perfectly preserved specimens came from different places on the Siberian northern coast; as that coastline reaches nearly halfway round the world, clearly, different spots on that coastline must have experienced different conditions at any one time. So where one set of mammoths was subject to dire cold, another might have been tasting tropical conditions.

---

1   It must be remembered that dating methods need a degree of licence, either side of the date stated.

That these repeated freezings have hit northern Siberia with what must be exceptionally low temperatures, and often in the summer – as indicated by what has been found in the woolly mammoths' mouths – points to that region becoming, if only temporarily, the equivalent of Antarctica in winter. The mammoth is not an Arctic creature. Ivan Sanderson, the naturalist, states:

> It now transpires, from several studies, that mammoths, though covered in a thick underwool and a long overcoat – and in some cases having quite a layer of fat – were not especially designed for arctic conditions; a little further consideration will make it plain that they did not exist under such conditions. That they did not live perpetually or even part of the year on the arctic tundra is really very obvious. First the average Indian elephant, which is a close relative of the mammoth and just about the same size, has to have several hundred pounds of food daily just to survive. Then, for more than six months of the year there is nothing for any such creature to eat on the tundra, and yet there were tens of thousands of mammoths . . . Therefore, the mammoths either made annual migrations north for the short summer, or the part of the earth where their corpses are found today was somewhere else, in warmer latitudes at the time of their death, or both.

It is hard to believe that the tundra would have been adequate for the massive amount of grazing required. Besides this, the rooted trees that have been found buried in close proximity to the mammoth remains are not to be found on the present-day Siberian tundra, which is well north of the treeline. That said, we have to remember that perfectly preserved remains date back for 50,000 years. This implies that the permafrost holding them has not, in some instances, been melted. So, although at times and in many places the northern Siberian climate has improved to a point where it can sustain large mammals, it has in other places not reached temperate levels.

It is difficult to check on the earlier periods when extinctions took place. However, radiocarbon dating indicates that around 15,000 years ago a change occurred. Conventional thinking says that the global climate began to warm up and ice sheets started to melt. Warming reached a peak about 13,000 years ago and then started to decline again, leading to a short period of renewed cold between about 11,000 and 10,000 years ago. This has been declared the true end of the 'ice age', and was followed by 7,000 years of relatively mild climate. What is fascinating as a pointer to very abrupt change is that about 13,000 years ago the global temperature is said to have soared by 11°F (6°C) within ten or twenty years – a far greater jolt than the current increases attributed to greenhouse warming. Obviously, it is our belief that this sharp turnaround only applied to specific parts of the globe.

**Other disasters have struck**

In the Tanana Valley in Alaska, whose river joins the Yukon, gold mining has taken place since the nineteenth century. Sluicing of tributaries of the Tanana has led to wide and long cuts being made in the stream beds. These have revealed an overburden of silt and muck that is full of broken trees and broken animal – mastodon, lion and mammoth bones have been found mixed with human artefacts – and all are in permanently frozen ground. F.C. Hibben of the University of New Mexico writes: 'Mammal remains are for the most part dismembered, even though some fragments still retain, in the frozen state, portions of ligaments, skin, hair and flesh. Twisted and torn trees are piled in splintered masses. At least four considerable layers of volcanic ash may be traced in these deposits.' The animals might have been killed by the volcanic eruptions, but their bones and the trees would not have been broken up in the manner described. This type of devastation points to storm action in the form of fast-moving wind or water. The destruction does not confine itself to central Alaska: similar damage is found on the Kuskokwim river flowing into the Bering Sea, and on the north coast of Alaska. If wind was the prime mover, it would call for a hurricane of force unknown to us today; if water, the sea would have had to flood the area with a force given only to gigantic tsunamis. All this appears to have happened recently – at the end of the 'ice age', or even later.

Caves and rock fissures are an interesting source of evidence for catastrophe. Throughout the world there are examples of caves filled with the bones of animals mingled together and invariably badly broken. These bones show little sign of being abraded by prolonged water submersion; their owners appear to have been subjected to sudden and immense violence which could only have come from an abrupt onslaught of wind or water – a tidal wave, for instance. Professor Buckland, geologist at Oxford in the early nineteenth century, describes a cave at Kirkland in Yorkshire in which teeth and bones of elephants, rhinoceroses, horses, deer, tigers (with teeth larger than those of the largest Bengal tiger) and hippopotami are found, dating to a period, in his opinion, after the last glaciation. Many of these animals had died 'before the first set, or milk teeth, had been shed'. Here we have animals that are not in the habit of fraternising so closely, and which all appeared to have died at the same time. Professor Joseph Prestwich, professor of geology at Oxford (1824–88) described fissures in certain rocks in the Plymouth area. These clefts, which were of varying widths in limestone formations, were filled with rock fragments that were angular and sharp (i.e. not smoothly rounded by the sea) and with animal bones that had been 'broken into innumerable fragments'. These bones were mammoth, hippopotamus, rhinoceros, horse, lion and numerous others. Many of these creatures are normally found in climates very different from present day south-west England. All skeletons had been dismembered, and the bones scattered, but they showed no signs of wear, and had not been chewed by other carnivores.

On the Mediterranean coast of France there are numerous clefts in the rocks crammed to overflowing with animal bones. Marcel de Seres in his survey of the Montagne de Pedemar wrote: 'It is in this limited area that the strange phenomenon has happened of the accumulation of a large quantity of bones of diverse animals in hollows or fissures.'

He found that all bones were broken into fragments, but not gnawed or rolled. No coprolites (hardened animal faeces) were found, indicating that the animals had not lived in these hollows. The same story is told by Prestwich about bone discoveries on Gibraltar: 'the remains of panther, lynx, caffir-cat, hyena, wolf, bear, rhinoceros, horse, wild boar, deer, ibex, ox and rabbit have been found in these ossiferous fissures. The bones are most likely broken into thousands of fragments – none are worn or rolled, nor are any of them gnawed.' He adds: 'A great and common danger, such as a great flood, alone could have driven together the animals of the plains, and of the crags and caves.' Among the bones on the Rock were found flints worked by Palaeolithic man, as well as broken pieces of pottery of Neolithic man.

The same story goes for Corsica, Sardinia and Sicily: numerous fissures in the rocks contain broken animal bones. The hills around Palermo have revealed extraordinary quantities of hippopotamus bones: twenty tons of these bones were shipped from one cave at San Ciro, and used for charcoal in a sugar factory. Much further away, in the village of Choukoutien in northern China, similar quantities of broken animal bones have been found with human skeletal remains. Each of the cave-fissure accounts has pointers in common: they show animals that have been violently buffeted and rolled over to the detriment of their bones. They show animals that have ended their lives prematurely. All this seems to suggest a violent surge of wind or water (more emphasis on the latter) which has flooded the area with alarming speed, killed and swept the creatures before it, and deposited them in any receptive caves and crevices. This water has then retreated in a comparatively short time –

short enough not to abrade the bones, and also in some instances to leave tissue in place. Then, too, there is a different type of marker: many of the animals are known inhabitants of far warmer climates than those provided by the places where they were found.

In the asphalt pits of La Brea on the edge of Wilshire Boulevard in Los Angeles, we have a parallel pointer to catastrophe. These pits are part of a much larger bituminous area that runs from northern California to Los Angeles. The tar is mixed with shale and sand and in places is up to 2,000 feet deep: the Pleistocene stratum is usually the top 150 feet of this formation. Animal bones are found in this formation in profusion. Perhaps the most spectacular among these is the sabre-toothed tiger, with its ten-inch-long canine teeth; it is found in large numbers, but with no intact skeletons. J.C. Merriam from the University of California writes: 'a bed of bones was encountered in which the number of sabre-tooth and wolf skulls together averaged twenty per cubic yard.' Among other animals found in this pit were bison, horses, camels, sloths, mammoths, mastodons and birds.

Now, where this catastrophe differs from the cave disasters, described above, is in the presence of tar. Animals swept into this tar should be found with their skeletal remains intact, bound by the hardened asphalt. It is not like that: the creatures have been severely battered and mostly dismembered before ending up in the pits which have then held them fast. It is interesting how many animals found there are no longer represented in North America. Judging by the numbers, there must have been many more animals living there at the time of this catastrophe than when Europeans first reached the west coast of America. In the pit at La Brea were also found the bones of a human – identical to those of Indians living in the region nowadays – who had met his end at the same time as the animals.

**Analysis of these deaths**

Analysing these disasters that have destroyed animal life in such an obviously violent way is difficult. Extreme cold, volcanic outbursts, violent winds and fast-moving water have all been contributors. The freezing of the mammoths on at least five occasions – probably many more – is a fascinating starting point to any analysis. As we have said, the level of low temperature was such that only a winter in the opposite hemisphere could equal it. We know that many of the dead mammoths were experiencing summer conditions when they died: flowers were found in stomachs and mouths.

So we are faced with astonishing evidence that says the mild-temperatured plains (in general these dead animals are found on flat lands that are low-lying, just above sea level) suddenly moved from a warm, congenial zone, in whichever hemisphere was enjoying its then summer, passed through the polar region of the opposite hemisphere, and ended up in its ridiculously cold vicinity. It would be inadequte to suggest that these mammoth feeding plains simply moved towards the pole in their summer hemisphere, despite the more obvious proximity, since a summer Arctic world would not have provided a low enough temperature to facilitate the required level of freezing. We are justified in assuming that the mammoths had been grazing on a latitude some 1,500 to 2,000 miles nearer the equator than they are found at the moment. So, suddenly – and that means almost instantly, within a day or two – the Earth would have moved on its axis through approximately 140°, to put these creatures at the opposite Pole.[2] This type of abrupt climate change seems to have happened a number of times along the north coast of Russia/Siberia, since carbon-dated mammoths at different locations date back to 50,000 BP.

It needs little imagination to visualise the happenings that would accompany such an abrupt axial move. Violent winds and surging of the sea onto the continental shelves would lead to the destruction of mammal life that occurred in Alaska and most other parts of the world. The fierce movement would have had a startling effect on the magmatic interior of the Earth: volcanoes that had been inactive would have sprung alive. The layers of volcanic ash mingled with the crushed bodies of animals supports this thinking. In a later section we will enlarge on the way this violence comes about.

**Land disrupted**

Disasters that build mountains do not appear to occur quite so rapidly as those we have just looked at. Or do they? If we look at the Altiplano – the high plateau between the two ranges of the Cordilleras in the Andes,

---

[2] Conversely, of course, if these mammoths had been grazing during summer in the southern hemisphere, they would have been abruptly switched to winter at the North Pole.

we see that it houses South America's largest lake, Lake Titicaca. This expanse of water was once many times larger than it is now, with an old shoreline which shows that it has been tilted, and that at one time the level was in places 350 feet higher than now. The average added height of the lake's surface was 90 feet above the present level. Something tipped that lake and spilt much of it. The former shorelines, although at different heights due to this tilting are comparatively modern and of one date – there do not seem to be any intermediate waterlines.

Curiously related to this anomaly is the presence of the ruined megalithic town of Tiahuanaco, which covered a considerable area at the southern end of Titicaca. This town was originally on the lakeside: it exhibits the ruins of an ancient port with evident quays. Now the town is many miles from the edge of the present, much smaller, Lake Titicaca. It is at 12,500 feet where corn will not ripen, and there are formerly cultivated terraces going up for another 2,500 feet.

Sir Clements Markham, in 1920, wrote: 'such a region is only capable of sustaining a scanty population of hardy mountaineers and labourers.' Apparently, the Cordilleras have risen several thousand feet since the town was built, and that was undoubtedly in the postglacial period. In fact, the uplift was even greater: Alexander Agassiz in 1875 identified contemporary species of marine molluscs in the lake, which suggests that the region had been previously at sea level. No one suggests that Lake Titicaca was elevated in a night, but it did happen rapidly. The tilting could have occurred even more so.

Until comparatively recently it was believed that the Himalayas had risen to their present height – Mount Everest being 29,002 feet – many millions of years ago. Not so now: marine remains of crabs, molluscs and fish have been found at the highest levels. In the opinion of Arnold Hein, 'these deposits contain palaeolithic fossils.' This suggests that these mountains have risen to their present levels within the age of man, conceivably within the last twenty thousand years. In Kashmir are to be found sedimentary deposits of an ancient sea bottom that was elevated to an altitude of 5,000 feet or more and tilted at an angle of 40°: the lake bed was dragged up by the rise of the adjacent mountain. Again, palaeolithic fossils were found in the lake's sedimentary bottom.

The Siwalik Hills on the south side of the Himalayas are a fascinating depository of animal remains. The hills have risen, since the last glaciation, to 2,000–3,000 feet. Amongst the mammal skeletons are thirty different variations of the elephant; all but one are now extinct. A.R. Wallace – competing with Darwin for the right to be deemed the inventor of the natural-selection theory – was the first to draw attention to the extinctions that had occurred on more than one occasion in these hills. That violent death had been meted out is clear from the number of broken tree trunks associated with the animal remains, and buried in the same alluvium. According to D.M. Wadia, the Siwalik deposits in a large number of different isolated basins have a remarkable homogeneity, so much so that an agent must have carried these animals and deposited them at the base of the Himalayas, on more than one occasion and over a span of many hundreds of miles. Violent wind and rushing water are the two most obvious natural villains.

**Volcanoes and earthquakes**

Generally, volcanoes come within the ambit of uniformity, but is this the way it should be? In the last two centuries, 250,000 people have been killed by volcanoes. However, 70% of those deaths have been attributed to four eruptions. The largest of these was Tambora, on the island of Sumbawa in Indonesia: in 1815 this 14,000 ft mountain blew off the last 4,500 feet of its cone in a massive explosion which expelled 50 cubic kilometres of magma in the form of hot fluids and pyroclastic flow. About 10,000 people were killed by the explosion and the tsunamis following the huge pyroclastic flows into the sea. Another 80,000 died from starvation caused by the ash deposited on agricultural land.

The next largest eruption came from Krakatoa in 1883. Again, it was an island in Indonesia, but this time only 18 cubic kilometres of magma were blasted more than 25 kilometres into the upper atmosphere. Twenty square kilometres of the island disappeared. Thirty-five thousand people were drowned by the resulting tsunamis hitting the adjacent shores of Java and Sumatra. In 1902, Mount Pelee on the island of Martinique discharged nearly one cubic kilometre of magma, in the form of hot gases and pyroclastic matter that swept down a valley and destroyed the port of St Pierre. Within minutes, 28,000 people were burnt to death. In 1985, the smallest of these four volcanoes, Nevado del Ruiz in Colombia, threw out several million cubic metres of pyroclastic material which landed on the ice cap surrounding the summit. The resulting meltwater, flowing down two sides of the mountain in a rapid mudflow, inundated two towns. Twenty-two thousand inhabitants were buried.

These were four minor catastrophes within 200 years; so how many larger catastrophes will have occurred during the last 20,000 years? The explosion of Santorini, 2,500 years ago, we have mentioned: it was much greater than Tambora and may have directly killed many more people. The fact that the caldera is some sixteen kilometres in diameter gives an idea of the volume of magma it once held. But these volcanoes are childlike compared with earlier eruptions.

During the last two million years the Yellowstone National Park area has undergone three major caldera collapses involving pyroclastic eruptions of anywhere between 300 and 3,000 cubic kilometres per explosion. The volume of ash alone from the latter, driven high and suspended in the upper atmosphere, would have been enough to block out sunlight round the globe for several years, leading to failure of crops and other plants. The starvation of man and beast would have occurred worldwide. After all, tiny Krakatoa caused disastrous harvests for two or three years in Southeast Asia.

Earthquakes, too, have been responsible for a vast slaughter in the last 500 years. In 1556 one earthquake in Shaanxi, China killed almost a million people. The 1737 Calcutta earthquake took 300,000 lives. The Lisbon quake in 1755 killed 60,000 people. This century, 180,000 died in Kansu, China in 1920; 143,000 in Tokyo in 1923; and once again China lost 750,000 inhabitants of Tangshan in 1976, with a Richter scale measurement of 8.2.

These figures do not say enough about the size in Richterscale terms of these earthquakes; they point more to the density of populations and the flimsiness of their housing. These quakes were undoubtedly large, but they were nothing like the size of those responsible for the upthrust of the Andes. According to R.T. Chamberlain, 'hundreds if not thousands of cubic miles of the earth, almost instantaneously heaved upwards, produced a violent earthquake which spread . . . throughout the entire globe. Many world-shaking earthquakes must have been by-products of the rise of the Sierras'. This goes too for the rise of the Rockies, Himalayas, Alps and more.

Quake activity does not confine itself to the land. In the 1923 Tokyo earthquake, the O-shima island volcano started erupting in Sagami Bay. Immediately this was followed by the floor of the bay groaning and twisting in a clockwise direction: its north shore moved some ten feet south-east, and the island of O-shima moved twelve feet north-east. Astonishingly, seabed soundings showed that the north end of the bay had risen 1,500 feet, and a mile further south the seabed had sunk by 2,400 feet. In the process a new island, 30 miles long, appeared; and one of Japan's jewels, the island of Enoshima, disappeared under the waves. In Agadir, too, a similar event occurred: the 1960 earthquake destroyed the town, killing 12,000 inhabitants, and then proceeded to lower the seafloor off the coast by 2,000 feet.

These events are the converse of mountain building, but they give a pointer to the speed with which great areas of the lithosphere can twist or move up and down. Anyone who has sailed along the Lycian coast of Turkey, for example, will have seen the movement that has occurred in the lofty cliffs west of Antalya. Here hundreds of feet of rocky stratification have been twisted in a moment from lying horizontally to lying near vertically; the sedimentary lines are almost unbroken, indicating that time has not been allowed to grind these lines to powder. The giant hand that has distorted the cliff faces has done it in a single abrupt movement, possibly overnight.

That earthquakes have not appeared catastrophic within recorded memory has contributed to the theory of uniformity. Nevertheless, we have to realise from the speed of mountain building, and seabed sinking, that earth movements do not always come under this heading. The pointers within the last 20,000 years are to events greater and more devastating than anything we are used to at the present. That they don't happen every century is obvious. But to say they don't happen every 5,000 years is harder. We are going to look at the trigger mechanisms by which catastrophes may have been initiated – and by this we don't simply mean marginally more serious earthquakes. We mean disasters that have wiped out great swathes of population, and even entire civilisations.

## Changing pattern of catastrophe

We have looked at cataclysms that have struck the animal world; we have seen also the dramatic geological upheavals that have recurred during the same period. However, the worst of these disasters appear to have taken place in the middle period, immediately after the alleged last ice age.

During the more recent part of the Holocene, disasters seem to have been on a lesser scale. For instance, volcanic activity is today much reduced: the world has thousands of dormant or extinct volcanoes, which were bubbling away during the earlier Holocene epoch. Now worldwide there are only some 500 active volcanoes. Iceland has over a hundred extinct volcanoes of geologically recent origin, and a far smaller number of active ones. Central America abounds in volcanoes that are no longer alive; the USA has few active, though there are many that have been at work in the recent past.

Hawaii has only the two mighty mountains of Mauna Loa and Kilauea, the largest active volcanoes in the world, whereas it has a dozen extinct ones. The Japanese islands possess droves of volcanoes that have been erupting in the not-too-distant past, and only a few that are still on the move. All round the Pacific Rim we see the ratio of recently extinct volcanoes to active ones to be high. Alaska, Kamchatka, the Philippines, northern New Zealand and the Pacific archipelagos tell the same story: what was highly active, magmatically, three thousand years ago is now quiescent. At the present rate of decline, there should be very few active volcanoes by the end of the next five hundred years.

# *What Caused this Earlier Violence?*

**The Earth's unstable axis**

There is enough evidence to conclude that in the early part of the postglacial a number of disasters occurred: death of the mammoths and destruction of wildlife on a massive scale, prodigious mountain building, and volcanic violence at levels very different from now. The last two thousand years does not seem to show such violence, and the conclusion is, if only from the volcanoes alone, that the activity below the Earth's lithosphere is quietening. So what was it that caused much more turbulence in the past?

This brings us to the central theme of this book. There is good reason to believe that throughout the life of the Earth there have been many shifts in its axis of rotation; and during the last 20,000 years – particularly since the end of the last 'ice age' – two or three more. In other words, the Earth has at times changed not only the angle at which the North–South Poles spin in relation to the plane of its orbit, but far more dramatically it has changed their locations: this means that axis no longer breaks surface at the present north–south polar points.

This view was examined in the nineteenth century by Sir George Darwin, the distinguished son of the more acclaimed Charles Darwin, and by the scientist Sir William Thompson, alias Lord Kelvin. Both felt there could have been axis movement, but not by great angular leaps at any one time. Kelvin admitted, however, that a larger shifting of the poles would have been possible if the Earth had a solid nucleus in the interior separated by a liquid layer from the outer crust. This he regarded as unlikely, directing his arguments towards an Earth with a molten interior. It is intriguing that the present conventional view takes the stances which he thought would permit pole movement – namely a solid interior wrapped in a liquid mantle – yet at the same time does not allow pole movement, because the mathematicians say 'no'.

An article in the *Geological Magazine* of 1875 puts their position emphatically:

> Mathematicians may seem to geologists almost churlish in their unwillingness to admit a change in the Earth's axis. Geologists scarcely know how much is involved in what they ask. They do not seem to realise the vastness of the Earth's size, or the enormous quantity of her motion. When a mass is in rotation about an axis, it cannot be made to rotate about a new one except by external force. Internal changes cannot alter the axis, only the distribution of the matter . . . about it. If the mass began to revolve about a new axis, every particle would begin to move in a new direction. What is to cause this? . . . Where is the force that could deflect every portion of it, and every particle of the Earth into a new direction of motion?

This adds up to the belief that the energy required to overcome the stabilising effect of the equatorial bulge[3] would be unimaginably large.

---

3    The equatorial bulge constitutes an expansion of the Earth – like many a middle-aged abdomen – round the equator. This means that the

In spite of this, more recently, Professor S.K. Runcorn, writing in Cambridge, has stated that polar shifts have taken place. He declares 'there seems no doubt that the Earth's magnetic field is tied up with the rotation of the planet in some way. And this leads to a remarkable finding about the Earth's rotation itself.' The unavoidable conclusion, Runcorn says, is that 'The Earth's axis of rotation has changed also. In other words, the planet has rolled about, changing the location of its geographical poles.' The conventional view has chosen to ignore this possibility, as we have said, on the grounds that the Earth's kinetic energy, gyroscopically supported by the equatorial bulge, would prevent the Earth from rotating other than in its equatorial plane. We are going to show that shifting the Earth's axis is not as much of a problem as has been made out. Then, too – as explained in Part II – because the Earth is being perpetually rotated by the solar wind, its innate momentum is not nearly as great as mathematicians would have it. Anyway, who really believes that the Earth's equator was always leaning over at 23.5° to the plane of the ecliptic.

Whichever theory of the solar system's origin is adopted, there is little room for planets to have started life with their axes other than at right angles to the ecliptic.

Therefore, as all planets rotate with different inclinations, there must have been movement in the

---

Earth is not totally spheroid: it is oblate, making the polar diameter 27 miles less than that of the equator. This bulge is given all the credit for gyroscopically stabilising the Earth's rotation. Without it the Earth could wander all over the place – its poles pointing in any direction.

angle of the poles at least – and that could have taken place long after the birth of these solar satellites.

An interesting position on the power of the equatorial bulge was taken by Dr Thomas Gold, when at the Royal Greenwich Observatory in 1960. Writing in *Nature*, he stated that if the Earth's interior is plastic – and we know that it largely is – the equatorial bulge is not that great a stabiliser, or a help in preventing capsizing. It must be remembered that because of this plasticity the bulge exists as a by-product of the Earth's spin. The absence of a bulge would create an untidy situation: although the Earth would continue to rotate in the same direction, in that it would maintain the same directional spin on its axis, that axis could be quite capable of pointing in many directions without interfering – other than minimally – with the Earth's momentum. An example of this is a beach ball that may be spinning on the surface of the water: although the axis will be spinning in one direction, the poles of that axis may move about and point in any direction, with only a fractional external force being applied to the ball. The converse is of course a flywheel, where the mass, instead of being spread over the entire volume of a globe, is largely concentrated in the 'equatorial bulge' of the flywheel's outer rim. The resultant gyroscopic effect makes it very hard to alter the direction of the flywheel's axis.

The consequences of the Earth's axis changing are of course devastating. The faster the change, the more disturbing would be the effect on wind and water. The degree to which the poles moved would also have a significant effect in the first instance on the stratospheric wind, where the highaltitude jet streams would encounter the solar wind at a different angle. This collision would be transmitted downwards, first to the lower wind movements, and then to the sea which would flood all shoreline areas, where the bulk of human population is to be found. Current groundlevel wind speeds, whose recorded highs have been known to exceed 300 mph, would be as gentle breezes alongside velocities occasioned by a change in the Earth's axis, where speeds in excess of 500 mph could be expected – inflicting the damage on animal life described above.

**The Sahara Desert**

Before moving to the mechanics of pole shift, it's worth our while taking a look at the Sahara. Prior to the ending of the last 'ice age', this vast region of northern Africa is said to have been an arid area, though confirmation of this is hard to achieve. However, some 11,500 years ago, the Sahara climate changed fairly abruptly, and the region became fertile. Thenceforth, it became the home of large herds of sheep, cattle and humans: it was populated with Europoid farmers in the north and Negroid herdsman in the south. It probably had a larger concentration of people than any other part of the world during that period. Then, with astonishing speed, some 4,000 years ago all that changed. Desertification started to set in – and rapidly. By Roman times the Sahara was deemed a land cursed by the gods.

In a fluctuating way the Sahara has shown evidence of inhabitants over the last 500,000 years: pebble tools in the Tenere are clear evidence of a lower Palaeolithic presence.

The old dried-out riverbed of the Tafessasset – at times hard to detect – shows a former major arterial waterway draining the Tassili, Hoggar and Air mountains through a 750-mile plain leading to Lake Chad. Historically this was home to many hominids: *Atlanthropus* (named by Professor Arambourg of the Muséum Nationale d'Histoire Naturelle in Paris, and deemed similar to *Sinanthropus* of Peking), Neanderthaloids (thought to have come from Europe at a later date during the last 'ice age') and *Homo sapiens* in the form of a Neolithic culture. The area has revealed remains of elephant, rhinoceros, hippopotamus, sabre-toothed tiger, giant warthog and lion.

Attempts have been made to link the climate of the 'ice ages' in Europe to that of the Sahara. A cold Europe was expected to invite a pluvial period on the Sahara; but this theory has not worked: there have been fewer oscillations of climate in the Sahara, according to received thinking. A temperate climate developed after the end of the last 'ice age'. Savannah-like conditions, not unlike southern Europe allowed cultivation and herding to develop over the next 7,500 years. Lakes and rivers were full of fish. Trees such as olives, cypresses and pistachios are still evident in the mountains. These attractive conditions led to successive waves of invaders – from Europe, from the south, and Ethiopian types from the east. The region is famous for its rock carvings and paintings, which show remarkable artistic development and reveal the progress from hunter-gatherers and fishermen to farmers. Pastoralists appeared with sheep and goats, and then cattle. The region had an economic richness that allowed specialist artisans to appear: jewellery has been found; special stonecutting became an industry; decorated pottery was evident over wide areas. What is interesting about the information gathered from the fantastic numbers of rock paintings is that it reveals the sequence of events, since the paintings are often overlapping one another.

At the height of the pastoralist culture trees such as the Aleppo pine, nettle tree, hollow ash, lentiscus, maple, alder and lime show a flora that is Mediterranean. Palynology (the study of fossilised pollen grains) tells us that from about 2,000 BC there began a progressive replacement of Mediterranean vegetation by one that is today characteristic of a much hotter and drier setting – possibly typical of one that is 1,500–2,000 miles nearer the equator.

What makes a desert out of a previously verdant land? The answer, in its simplest form, is that evaporation exceeds rainfall. The amount of rainfall allowed by this definition of the word 'desert' is less than ten inches per year; the Sahara easily fulfils that criterion and as such is the world's largest desert. It covers 3.32 million square miles: its nearest rival in Arabia covers less than a third of that area. This drying out, which started around four thousand years ago, had dramatically changed the region within a thousand years of that start: greater heat and less rain had become the pattern for this vast area. With less vegetation for grazing there was a tendency by the Tuareg's predecessors to cut down what trees remained, thus drying out the land even further. This dreadful habit is still contributing to the enlargement of the desert. There have been attempts to claim that the Sahara was once part of the sea, but this argument is not sustainable. The African shoreline has changed little over the last million years. There is clear evidence that prior to 4,000 years ago there was little sand in the region, other than some large dunes in the south; and the place was heavily populated. Also, the granular structure of the sand is of a non-marine-generated type. So this is a desert that comes about because otherwise fertile land has not received, or been unable to retain, moisture.

Something dramatic happened some 3,500–4,000 years ago. We know that hundreds of mammoths died at that time – quick-frozen on Wrangel Island, just north of the fareastern Siberian coast. As we have said, these animals died with summer feeding in their stomachs, but under arctic winter conditions. Summer grazing for mammoths would have been in temperate latitudes, some 2,000 miles from the equator. We believe these creatures were grazing in the southern hemisphere, when a polar-axis shift occurred: this took them through 140° of latitude, via the North Pole, to the present location of Wrangel Island. They were deepfrozen as this pole shift took them through the Arctic winter.

Such a polar movement had a similar effect on the Sahara: terrain which would have been situated in the southern hemisphere, with a climate comparable to South Africa, would have moved abruptly to the northern hemisphere at a latitude much closer to what became the new equator. In this case, a congenial rainfall would have instantly disappeared, to be replaced with the start of desert conditions. The distance moved by the Sahara would not have been the same as the distance moved by the mammoths. Such a pole movement calls for the Earth to tilt, pivoting on two points situated at the equator; where those two points are determines how far a particular region will move in relation to the plane of the ecliptic.

It is interesting to note that at the same time as the two above events occurred, the Great Basin area of North America – covering the desert parts of Nevada, Utah and Arizona – also changed its climate. From being a well-populated region, it commenced some 4,000 years ago to become the desert it is today. These changes are contemporaneous with catastrophic events recorded in written history, which we will examine in a later section.

## *Pole Shift*

### The Earth's axis has many choices

We believe the poles have changed abruptly, moving huge distances. As previously stated, there are two fundamental types of axis change. The first is when the existing axis changes its angle in relation to the plane of the ecliptic: this means, of course, that rotation continues with the existing poles uninterruptedly, but the angle of tilt causes the seasons to change more rapidly, or more slowly. The second is when the Earth develops a totally new north–south polar axis. It is pretty evident that the two types of change are interdependent – it is certainly not possible to change to a new axis without first turning the old axis through anything up to 180°, because only then would the solar wind (as described in Part II covering the New Theory) responsible for the rotation be able to stop that spin and restart it using a new axis.

Incidentally, this is what has happened to the planet Venus. It has fallen over so that its equator is at 177° to the plane of the ecliptic. It is now rotating very slowly, since the solar wind is no longer brushing the surface and causing the previous rotation: soon the solar wind will cause rotation to cease, prior to reversing the rotation in a prograde direction. In Venus' case this cessation will take somewhat longer than it would on Earth, due to the thickness of Venus' atmosphere. Solar radiation does not reach that planet's surface so readily.

Let's examine some of the possible polar positions available. What better start than the Earth's present one. The Earth wobbles about in a seemingly drunken way to begin with; every 25,800 years the poles sweep out the 360° of the zodiac (precession of the equinoxes), like a top slowing down; the North and South Poles each trace out the base of a cone in space with their apexes at the Earth's centre. At present the North Pole points to Polaris in the Little Bear Constellation, but four thousand years ago Alpha Draconis was the star which appeared nearest to the pole. Then, too, this axis point goes through a 41,000-year cycle during which it varies the angle that the equator holds with the ecliptic between 21°26' and 24°36'. And further, every 93,000 years the average distance to the Sun goes through a cycle which changes the Earth's orbit from circular to elliptical and back again – switching the radius of the Earth–Sun by three million miles from its average distance – which in its turn alters the celestial direction of the poles.

Looking at all this wobbling through the eyes of the New Theory, as described in Part II, it is doubtful whether the Earth has ever run the full course of one of these lengthy cycles. It's far more likely that a pole shift has radically altered the Earth's rotation, making one or all these cycles meaningless. To start with, the solar wind, in rotating the Earth, has a greater impact on each hemisphere at different times: the northern hemisphere is brushed more in its summer, the southern hemisphere more in the northern winter – and vice versa. This rotational propulsion by the solar wind is very uneven, and calculated to infect the Earth with a wobble. If that wobble is not too apparent (or is blamed on one of the other reported wobbles), it is because the amount of energy applied to each hemisphere is the same. That said, this treatment could be dangerously destabilising because the two hemispheres do not have the same amount of albedo – ability to reflect sunlight – one has more land, the other more ocean, making the reflectivity different in each half.

Attention must be drawn to the effectiveness of the solar wind in its rotational work. The more the Earth's axis is tilted, the less efficiently the solar wind strikes the

Earth's surface, and the slower goes the Earth's rotation. If by chance the polar axis were to be perpendicular to the ecliptic, there would be far greater rotational speed, because the net solar wind would be tangentially brushing the surface of the planet, parallel to the direction of rotation. The wretched feature of this angle is that seasons cease to exist – summer and winter are the same – and it's pretty dreary at the poles, which never get beyond equinoctial temperatures.

A very different angle of tilt would be if the Earth fell over some 70°, and lay with its spin axis parallel to the ecliptic. Before long, rotation would cease, giving the Earth a chance to establish new poles. After this the solar wind would restart rotation based on the new axis. This is what is happening at the present time to Uranus: its equator lies at 97° to the plane of the ecliptic,[4] but is still rotating approximately every sixteen hours. It will gradually slow down, being unsupported by the solar wind which is no longer brushing the surface in concert with the direction of spin. It will come to a halt. Slowly, the solar wind will develop a new rotation for that planet. The new poles will be nearly at right angles to the ecliptic: they will be found at the two points furthest from the ecliptic when Uranus has ceased rotating.

Let's look, for a moment, at what follows an axis shift of this sort. Once the Earth has moved the direction of its equatorial plane, the solar wind is no longer urging rotation of the planet by brushing that equator. As the

---

[4] A tilt of 97° is very extreme and only goes to emphasise that this planet did not start life like this – what this book is all about. That it is also rotating so fast compounds this argument.

Earth slows down and stops, the equatorial bulge will cease to distend, and the Earth's oblate shape will become more spherical. As soon as a new rotation, and hence a new equator, has been started, a fresh bulge begins to expand, as the Earth's plasticity allows. This calls for a considerable movement of magma within the mantle: there will be a general shuffle in the direction of where the new bulge will grow. We could then expect to see a massive increase in volcanic activity. All those thousands of dormant volcanoes worldwide would start to sit up, and join in the prolific spewing of magma.

Someone is bound to ask, at this point, how the next obliquity of the axis develops. Obviously, the life of the new axis will begin with the solar wind brushing the new equator parallel to its own flow. So there will be no seasons. It is during this early period of very slow rotation when instability is greatest, and the new equatorial bulge has not yet developed, that a tilt is likely to occur. Later, as the magnetic field grows,[5] and stability sets in, the Earth enjoys a period of peaceful rotation, before again encountering the disastrous events that this book is about.

Perhaps the most likely shift – one for which there seems to be support in the palaeomagnetic evidence produced by cooling lava flows and some sedimentary rocks (see section on Earth's magnetic field) – is one of 180° or just short of this angle. The Earth turns completely over, and North and South Poles change places in relation

---

[5] Received science believes that the Earth's magnetic field is created by movement inside the Earth coupled with the Earth's rotation. Once the Earth has changed its axis, there is immense internal movement, provided there is rotation. So until there is movement internally joining with global rotation, there is no magnetic field.

to the ecliptic. At that time the Earth is exhibiting just the same momentum as before the shift. It continues to rotate as before, but the two poles are now on different sides of the ecliptic. The Sun still rises in the east and sets in the west; however, the firmament has changed: the North Pole points to the sky which is presently over the South Pole. The solar wind – in accordance with the New Theory in Part II – will now be offering far greater opposition to the Earth's rotation than it did when the Earth might have only fallen some 70° and be lying with its axis parallel to the ecliptic. As before, very gradually the Earth's rotation will slow down, come to a stop, and restart in the opposite direction. The Sun will then rise in the west and set in the east. And this is something that history has already reported, as we will shortly see.

**Another choice of pole movement**

Another theory that has found greater favour in that it does not receive as much hostility from the mathematicians, is the belief that the Earth's crust has slipped in the past. This takes the poles away from their present positions, but leaves the core, mantle and lower lithosphere where they presently are. The ratio of the Earth's crust to its interior is far smaller than an eggshell to the egg's interior, so little momentum resides in the crust. This theory was offered in the nineteenth century by J.H. Pratt, and it was supported in the twentieth century by the astronomer Arthur Eddington, who attributed the ice ages to the shifting of the Earth's outer crust over its interior. The problems with this theory are deep: firstly, isostasy – the understanding that mountains, rather like icebergs, are floating in a viscous liquid magma with a comparable mountainous volume under the surface of this magma – means that the underside of the crust is seriously uneven. The distance to the mantle under the oceans is much less than under the continents, so, with these downward protuberances, shifting the crust would be no smooth slide. Then, too, to slide the crust over the equatorial bulge would call for not only further force, but an elasticity in the crust that is hard to believe. The bulge, with its extra 27 miles, would be one hell of a Beecher's Brook for the sliding crust to jump. Secondly, the force required to move this friction-held crust would be comparable to that required to stop the Earth's primary spin. Two tight-fitting circular items take much force to move them, as we often find when trying to remove the lid of a jam pot; so moving two tight-fitting spherical surfaces becomes exponentially harder. No one has satisfactorily suggested where this force might come from: a close-passing asteroid would be hopelessly incapable of producing the right directional force which would have to be a torque applied to the entire surface.

Professor Charles Hapgood, a keen supporter of this idea, took the view that the thin but rigid outer crust could at times be readily displaced, moving in one piece 'over the soft inner body, much as the skin of an orange, if it were loose, might shift over the inner part of the orange all in one piece'. Albert Einstein investigated the possibility that the weight of the ice caps, which are not symmetrically distributed over North and South Poles, might cause such a shift. He wrote:

The Earth's rotation acts on these unsymmetrically deposited masses, and produces centrifugal momentum that is transmitted to the rigid crust of the Earth. The constantly increasing centrifugal momentum produced this way will, when it reaches a certain point, produce a movement of the Earth's crust over the Earth's body, and this will displace the polar regions towards the Equator.

Both these powerful thinkers ignore the above objections; they overlook the very severe pressures compounding the friction at the base of the lithosphere. Let us suppose that all the above objections to crustal movement are overcome: we are still left with the fundamental shape of the Earth. It is oblate – a sphere with flattening at the top and bottom. To move the crust, and find that it will fit the interior, however plastic, is asking too much. So this theory can be safely filed away. We must return to the question of how to move the globe.

No one single agent causes this pole shift: it's an interplay between the magnetic field of the Earth, the albedo of the Earth – particularly at the poles – and the solar wind/radiation. As we mentioned, the solar wind rotates the Earth in an uneven way, which is not destabilising provided the thrust is given equally to each region. Anyone who has seen a child beating a hoop along a path will recall how disconcerting it is for the hoop to receive clouts anywhere but evenly applied to its rim: without this, it wobbles out of control and falls over. The Earth is not dissimilar. As more ice melts at the

North Pole, at the South Pole the albedo increases, and the chances of the solar wind dishing out an even-handed thrust to each hemisphere become less. Here, too, is where Einstein's idea of the significance of symmetrical ice masses building up weight at the poles can complement the other agents. We will look in the next sections at the historical role attributed to ice, and the way the Earth's reducing magnetic field alters the picture.

## *There Never Was an Ice Age*

### The polar ice sheets simply moved with their poles

One of the first effects of adopting a theory of pole shift, is that it puts paid to the belief that there were ice ages in the accepted sense. By 'ice age' is meant a considerable increase in the volume of ice and the surface area of the Earth covered with ice. This implies a worldwide drop in the overall temperature of the Earth by at least 5°C, according to received science. Of course there are fluctuations in the world temperature, due to solar activity such as sunspots and solar storms, and also due to changes in the distance between Sun and Earth as the latter pursues its elliptical route round the Sun. There are also changes in temperature due to alterations in the relationship between sea currents and the land: for instance, a southward movement of the Gulf Stream could allow a significant build up of ice in the Southern Arctic Ocean, with a concomitant change of temperature. Changes in El Niño and La Niña in the South Pacific provide dramatic variations in humidity and temperature south of the equator. All the same, it is probable that the average energy output of the Sun has been constant over the last few hundred thousand years; so we must look to other reasons for the apparent ebb and flow of ice sheets in both hemispheres. Looking for explanation to the Earth's route round the Sun is not satisfactory. Generally speaking, solar output and solar distances, when added together, do not lead to an alteration in the world average temperatures of more than 1.5° to 2°.

At this point it is worth underlining the immense complication surrounding the arrival of the Sun's energy at the Earth. Here is a passage from John and Katherine Imbrie's *Ice Ages*, which shows how self-compounding, or the converse of that, an ice sheet can be:

> At only two latitudes is there an exact balance between these *energy* gains and losses: 40° North and 40° South. At all other latitudes the radiation budget does not balance, and therefore, at these places, there is a tendency for the Earth to heat up or to cool down. Near the equator, the imbalance tends to raise the temperature. Land and sea surfaces there absorb much of the incoming radiation, days are long, and the sun is high in the sky. Near the poles, on the other hand, there is a net loss of heat because the ice and snow that are present there reflect much of the Sun's energy. In addition, the Sun's angle is low at these latitudes. Unless processes other than reflection and radiation were at work, the poles would grow colder each year and the equator hotter. Winds and currents prevent this from

happening by transporting heat from the equator toward the poles.

Trade winds and hurricanes are examples of this heat transport mechanism, as are the Gulf Stream in the Atlantic and the Kuroshio Current in the Pacific. Simultaneously, the south-flowing currents along the eastern sides of the North Atlantic and North Pacific transport cold water towards the Equator.

Any valid theory of the ice ages must take into account that the growth or decay of a large ice sheet would have a large effect on other elements of the climate system. For example, if an ice sheet on land is to expand, water must be drawn from the oceans, carried through the atmosphere to the site of the ice sheet, and precipitated there as snow. Variations in the volume of global ice are therefore linked inextricably with variations in sea level. Furthermore, any change in the area of an ice sheet must bring about a change in the radiation balance of the globe. When an ice sheet expands, heat is lost through reflection, global temperatures drop, and more ice is formed. Conversely, when an ice sheet shrinks, temperatures rise, and further shrinkage occurs. This 'radiation-feedback effect' plays an important role in several theories of the ice ages because it explains how a small initial change in the size of the ice sheet is amplified. The main objective of most theories is to discover the cause of this initial change.

As has been said, a prime difficulty with generating ice on land is that paradoxically it calls for more heat, rather than less, at the equator. Only by causing more water to be lifted by wind and evaporation in the tropical regions does the Earth drive added moisture towards the Arctic and Antarctic, thereby allowing it to fall as snow. The snow then becomes impacted as ice caps and sheets. Thus, purely lowering temperatures worldwide does not create an ice age. Nevertheless, lowered temperatures lead to frozen seas, and these increase the albedo, keeping the temperatures in their reduced state. So once into a cold period, the Earth has difficulty shaking itself free and backtracking to former temperatures.

Another variation in the Earth's conduct which leads to enhanced cold is the angle of tilt that we mentioned in the last section. When the axial tilt moves towards 68°34' (the equator is tipped 21°26') there is little alteration to the tropical temperatures, but in the higher latitudes the summers and winters are colder: in summer each hemisphere in turn points less in the direction of the Sun, so the previous winter's build-up of ice does not get melted. This can lead to a net increase in ice volumes but it is not going to lead to the mysterious worldwide heat drop of 5°C which conventional science seeks. It's worth noting that 10,000 years ago, when we had just left the last 'ice age', the axial tilt was at its maximum (equator tipped 24.5°), which meant hot summers quite capable of destroying the previous winter's ice accumulation. But this argument is not strong enough to suggest that this tilt destroyed the 'ice age'. After all, the Earth had been approaching this angle long before the 'ice age' finished.

**Earlier thoughts on ice ages**

It was Milutin Milankovitch, a Yugoslavian mathematician, who earlier in the nineteenth century held centre stage with his astronomical theory of ice ages. He had painstakingly put together the three effects of the Earth's location and movement: the tilt of the axis, the change of the Earth's orbit from near circular to elliptical and back again, and the precession of the equinoxes. His calculations from these three varieties had taken him back for 600,000 years. Always his desire was to show that the impact of summer heating on the higher latitudes of 60°–70° was what caused an ice sheet to increase, or the opposite. According to T.F. Gaskell and Martin Morris, he accepted that there was often a "time lag between the time of maximum heat from the Sun and the temperature changes that occurred on Earth, as determined from biological and other evidence, because it takes several thousand years to diminish the ice age cover of snow and ice . . . on account of the high reflection of energy. If this change in the Earth's albedo is taken into account, it appears that the alternation of ice ages and warm present-day climates corresponds to a world average temperature variation of above 5°C."

Many scientists felt that Milankovitch's theory might account for 60% of 'ice ages'. However, there was a major problem. The eccentricity of the Earth's orbit is only 0.017: this meant that the radiation – reducing as the square of the increased distance from the Sun – would account for only 0.3% of the fluctuation in solar intensity. This was not good enough: an external explanation of ice ages now has to take second place to an internal explanation – greenhouse gases, according to received science.

The Little Ice Age – and extension of the Maunder Minimum – which lasted from 1450 to 1700 (or later, depending on interpretation) increased glacier sizes and ice cover in the northern hemisphere at least, and lowered the world's present average temperature by 1.5°C at its peak. It is unlikely that falls in world temperature have exceeded this figure, if caused by the numerous other theories contributed by astronomical and geophysical scientists. These theories include clouds of dust particles within the solar system which block off solar radiation arriving at the Earth's surface; and a fall in the level of carbon dioxide – the greenhouse gas which is present in very small volume (0.03%) in the atmosphere, but which is effective in permitting the passage of incoming short-wave radiation, yet is opaque to the reflection of long-wave radiation back into space. Then, too, there is volcanic dust from large eruptions which darken the atmosphere, preventing incoming radiation, and lowering world temperatures. In the nineteenth century Charles Lyell, the early geologist, even believed that ice ages were caused by an uplifting of the Earth's crust – a general raising of land heights leading to lower temperatures. More recently, Alex T. Wilson, a New Zealand scientist, has proposed that a large ice surge in Antarctica could result in a considerable part of the ice sheet sliding into the sea: this would add noticeably to the albedo and hence provide a cooling.

Another theory came from Maurice Ewing and William Donn in 1956. They suggested that present Arctic temperatures could lead to an ice age if more moisture were conveyed by warm North Atlantic currents, and made available to increase the snowfalls in that region. Hence, by increasing the ice-covered area, the albedo would lower temperatures. This is not a totally novel idea, but it underlines the importance of temperature being moved around the Earth by sea transport, a matter that the Gulf Stream makes very clear to north-west Europe. All these theories may well add their little bit to the changing climate of the Earth, whose component activities – wind, sea currents, glaciation, tectonic movement, vulcanism and a host of others – interlock in a way that defies computer modelling. Then, too, of late it is thought that one of the more telling causes of climate change has been fluctuations in the level of greenhouse gases. The scientific world has become far more focused on this cause since the awareness of man's contribution to the heat-lifting gases became a well-sung theme in the 1980s. Incidentally, the meteorological world of the 1960s had promised us a cold era starting about that time. So the sudden arrival of the greenhouse effect still carries question marks for many people.

It is understood that in Precambrian times the Earth's atmosphere had a 20% carbon dioxide content. The Sun was emitting less radiation, yet the Earth was warmer. The greenhouse effect with that amount of gas about would have kept the Earth warmer than at present: this, too, would explain why there were no 'ice ages' (little if any ice at the poles) during the first three billion years of the Earth's life. All that changed with the arrival of plant life: photosynthesis ultimately led to our atmospheric content of carbon dioxide being at its present low. Vast quantities of carbon dioxide have been locked up in the Earth's limestone rocks where the carbon part is 100,000 times more than in the atmosphere. Consequently, the Earth has lost considerable heat over the last billion years, in spite of the fact that the Sun now emits more radiation. The present belief is that variations in atmospheric carbon dioxide – and methane as well – occur regularly, and have a pronounced effect on temperature in the higher latitudes. So how does fluctuation occur? This is only possible when the carbon-bearing rocks have been subducted at the tectonic-plate margins, taken down into the mantle, melted and their carbon dioxide released through volcanic action.[6]

But as we have previously noted, the present volcanic activity is reducing; so the carbon dioxide presence should also be reducing. It is ironical that man is now releasing a plethora of greenhouses gases – so the net effect may be a slight increase in world temperatures. It does, however, show hubris on the part of man that he believes himself so powerful as to be capable of altering world temperatures.

More than all the other interlocking excuses for a change in world temperature, the Earth appears to be affected by fluctuations in these gases. That is the case

---

[6] Seafloor spreading from the mid-ocean ridges carries carbon dioxide locked into limestone on the seabed. This seafloor is then led down the various oceanic 'deeps', and back into the Earth's mantle, where it is melted, prior to recycling.

until we look again at pole shift as a cause of major temperature change. Fluctuation in the level of the gases should account for little more than worldwide swings of 1.5°C. This draws our attention abruptly back approximately 11,500 years to the end of the last 'ice age'. From 13,000 years ago, the temperature – according to the oxygen-isotope records extracted from ice cores taken from the centre of the Greenland ice sheet – changed rapidly to colder conditions.

Then fifteen hundred years later, with a speed that makes a nonsense of all but one theory on world temperatures, the cores from the same spot record a rise of 7°C within a period 50 years. This is a truly staggering change. None of the above theories could account for so sharp a swing: only a shifting of the Earth's axis will explain such a rapid alteration. It was at that time, therefore, that the North Pole moved from its then location in the Davis Strait on the Arctic Circle – and comparatively close to where the ice cores were collected – to its present site.

We must emphasise at this point, that pole shift can only be blamed for reported changes in world temperature when those changes are both large and rapid. A combination of some or all the other theories explains the lesser temperature changes. That the change should have been so accentuated and so fast 11,500 years ago compels us to take this theory seriously. It is also likely that the change happened in much less than fifty years. It must be remembered that the sudden reported increase in temperature is not a worldwide reading: it simply shows that a particular spot has surged in temperature as the point of extreme polar cold moved further away.

### Ice-age theory doesn't fit the evidence

But more of pole shift: changing the location of the poles means that ice sheets which generally speaking fill the Arctic and Antarctic Circles, appear in regions which are today much further south in the northern hemisphere, or much further north in the southern hemisphere. However, as there is little or no overall global increase in the area of ice coverage (that variation is dependent on the type of land that is glaciated: if mountainous there is an ice increase, if lowland, an ice sheet reduction), it takes different pole shifts to account for the presence of ice in places as far apart as, for instance, New Zealand, Patagonia and South Africa. All of these have been glaciated in the recent past. The original theory of ice ages sprang from the inventive mind of Louis Agassiz, 1807–1873, who believed that ice increased from the poles, spreading towards the equator. At the height of the last ice age, ice was thought to cover three times as much of the globe as it does now, and in the northern hemisphere alone glaciers covered approximately 10 million square miles, which is thirteen times the area covered today.

Well, this theory hit a snag. To the surprise of the nineteenth century geologists after Agassiz, there turned out to be a northern limit to the ice within the Arctic Circle as well as southern boundaries which were marked by the piles of moraine pushed to the extremities of the ice sheets and glaciers. In fact, ice sheets have spread out from different centres, and at different times separated by what have been described as interglacial periods. The Laurentide Ice Sheet, for example, grew and spread out from a centre which was near the Hudson Bay at latitude 60°N: from there it flowed north to the shores of the Arctic Ocean. This event indicates clearly that at that time, the Laurentide centre was the north pole of its day.

An interesting feature of climate variation comes from the cold centuries of the Little Ice Age, when the old Viking sailing route to Greenland became impassable as the Arctic sea ice spread southward along the eastern and south-west shores of that land. The ice at one time extended south of Iceland to the Faroes – even allowing a polar bear to reach them from the ice sheets – according to T.F. Gaskell and M. Morris. About this time, in the southern hemisphere, Captain Cook managed to go much further south than the present-day pack ice allows. This indicates reduced ice in one part, at least, of the south polar region at that time.

Further evidence from the rainfall records in South America seems to indicate that the climatic zones of the Earth drift south and north together, so that if the Arctic ice moves south towards Iceland, it recedes in the southern hemisphere. It is extremely hard to find an explanation for these seemingly opposed temperature movements in the two hemispheres.

First glances would suggest this as a fine example of pole shift. But it doesn't fulfil the requirements. For a change in the location of the poles, the Earth's rotation has to come to a standstill, as we explain in Part II, and we know this didn't happen. A change in the axial tilt does not give the correct answer: the temperatures in the two hemispheres would change in the same direction at the same time. At any rate, this type of tilt change really leads to a change in the length of the seasons, which in its turn produces greater extremes in the seasons at the higher latitudes – hotter summers, colder winters. The answer may lie in the sea: a change in direction of Atlantic currents would have startling effects on the polar regions, and these might well be different in each hemisphere. These temperature changes in the Little Ice Age were only by a degree or so – nothing like the extensive polar movements we believe to have occurred on several occasions since the beginning of the last accepted 'ice age'.

The fact that evidence abounds of different 'ice ages' in America and Europe must make us question the conventional theory. For example, in North America four main ice ages are accepted as having happened over the last 125,000 years, while in Britain and Northern Europe seven such glaciations are recognised. Surely, these facts alone point to a shift in the locations of the poles. They certainly do not point to a worldwide increase in glaciation. After all, the Laurentide Ice Sheet is said to have parted from North America some 2,000 years after the European ice sheet had melted.

**Sea levels**

In our search for clues that point to the pole shifts, we come upon the important question of what sea levels have been saying. Throughout the world there is evidence that the sea has been at many different heights from its present one. What is nice about this subject is the knowledge that sea levels, with minor contradictions, are the same all over the world. So, unlike winds, tides, currents, temperatures or continental/maritime effects, which vary at different places round the world, they can provide objective information that is harder to dispute. That data gives reasonable pointers to the worldwide volume of ice that is to be found on land, rather than floating in the sea. Having said that, we must be careful to look for signs that the land has not altered its elevation: when ice has lain on the land for a period, it isostatically presses the Earth's crust down into the mantle, thereby giving an incorrect sea level reading. Likewise, when the land is released from its overcoat of ice, it slowly rises to what would be its isostatic 'norm'. Sometimes this rise, while taking time to get started, is quite rapid: the land round the Great Lakes of North America, which was depressed during the last 'ice age', is at present rising at the rate of more than a centimetre every three years. Because of the problems created by ice pressure on the land, it is much better that sea levels should be assessed at places which have not been subject to the overburden of ice. That probably means in the lower latitudes, where the so-called last Pleistocene ice age is concerned.

The sea-level evidence is sharply etched on the hills round the coasts. The west coast of Scotland shows two or three tiers going up 400ft or more above the present sea level. Svalbard, north of Norway, where there is scant vegetation, shows similar steps in the hillside even more clearly. Many other coastlines worldwide mark a previous sea level with a raised flat plain stretching back from the shore. Equally, below the present sea level are submarine shelves at different levels down to similar depths or even deeper. The length that the shelf (above or below the surface) stretches indicates the length of time the sea remained at that level: once a new level has been established, wave action against the shore causes hunks of shoreline cliff to collapse into the sea, which then carries off the debris, leaving a flat sea bottom. It's worth observing how sharply defined these levels are: each plain may be separated by heights of twenty or thirty metres; and generally there do not appear to have been enduring sea levels in between them.

A point to note, which only concerns the conventional 'ice age' theory, is that when there is an interglacial, the warmer climate causes the warmer seas to expand in volume, thereby raising the sea levels yet further. If we accept the understanding that there never was an 'ice age', only changes in the polar locations, then there would not have been such worldwide extravagant changes of temperature, and the sea levels would not be affected by volume expansion. It must be remembered that sea levels are only lowered if the ice in the polar regions accumulates on the land: ice forming in the sea may actually raise the sea level (due to the increased volume of ice – measure for measure – over water). Ice forming on land arrives as snow which has travelled polewards from ocean water evaporating in warmer climates. While ice forming on the sea is, of course, frozen *in situ*. The departure of the water vapour from the temperate and equatorial zones has, naturally, lowered the ocean levels.

Here we should add that stretches of land separated by channels of water count as land: the ice builds up on each piece of land and joins over the water. Shallow marine basins fill with ice which then behaves as though on land. It is, in fact, the ice that builds above sea level, whether on land or on submerged ice, that contributes to the net lowering of worldwide sea levels. In Greenland and Antarctica there are seismic indications that both land masses are made up of more than one island, but the glaciation is such that each behaves as one land mass.

**Sea levels with pole shift**

How do these changed sea levels differ with pole shift or conventional 'ice age'? An 'ice age' would see worldwide water levels going much lower than they are at present, due to the sheer volume of frozen water locked up on land, above sea level. It has been suggested that if the area of the Earth's surface covered with ice was three times that of now, the sea level would have had to fall by some 80 metres, and that allows for much of the ice to be floating. As it happens, the averaged shorelines of the world do not appear to have gone down more than 35 metres during the last 20,000 years. That is wholly inadequate to explain the extra amount of ice allegedly generated in both hemispheres.

The amount of ice created by pole shift can vary, not only depending on the height of the land, but also on where the new poles locate themselves. For instance, if it were possible to locate both poles entirely on land – which at present is impossible, due to land not being available at 180° removal – the build-up of ice would account for the lowest historical sub-sea shorelines. These low sea levels were possibly achieved several million years ago when continental drift had located two land masses 180° apart which became the poles of the moment. Continental drift is depressingly slow, and may not always move in the same direction, according to Alfred Wegener, its originator: so it puts itself outside the range of these discussions. Theoretically, with pole shift there should be little change in sea levels.

In practice the ice surrounding the old poles melts more rapidly than the new ice forms at the new poles. Moving the poles to warmer spots results in the direct application of solar heat to the old poles, while the new poles depend for their snow on moist air coming from warmer regions. To this extent the sea level would rise for a short period, and then return to its pre-shift level. This reversion might take many years, but depending on the air circulation, could happen much more rapidly. Far more important for sea levels is where the new poles are stationed. If there was more land within the Arctic or Antarctic Circles at the last stopping point of the poles, the sea level would at that time have been lower. And that is just the way things were.

**Where were the last poles?**

Evidence points, as we have said, to one of the last north poles being in the Davis Straits between Greenland and Baffin Island, at approximately 66.5°N, on the line of the Arctic Circle. If this were the spot, glaciation in directions south and eastwards would extend the ice-sheet extremities to where we know they were at the end of the last 'ice age'. In Europe this sheet covered Britain and touched the north of France; it covered Scandinavia down to northern Germany; in Russia it extended nearly as far east as the Urals, though only touching the northern part of the country. Ice was not present in Siberia. In North America the ice sheet went as far south as Washington, covered the Great Lakes, and reached the Canadian Rockies, where it met up with glaciers coming from the higher ground.[7]

There was no ice in western Alaska. Isn't it a curious coincidence that these ultimate reaches of the last glaciation should be the same distance from the place where the Arctic Circle cuts the Davis Strait, at Cumberland Peninsula, as the latter place is from the North Pole! Then, too, this location would mean that far more land would have been glaciated than now. The ice sheet would have covered the large islands of the Canadian north-east, together with the shallow waters separating these islands and Hudson Bay, which for that period should be deemed land. (Most of the Arctic Ocean would have been free of ice, while a part of the north Atlantic would have been iced over between Greenland and Europe.)

In the southern hemisphere, the corresponding location 180° away from the Cumberland Peninsula is

---

7   There is some evidence that suggests the ice did not reach the Rockies, but left a passage running north–south between the Rockies and the ice sheet, through Alberta and Montana. Artifacts found in this passage have been carbonated to before the melting of the ice sheet.

a point on the edge of the Antarctic land mass at 120°E. This location would have meant that Western Australia was just outside the reaches of the ice; south-eastern Australia, Tasmania, and the southern island of New Zealand marginally in range.

According to John and Katherine Imbrie, these places were slightly glaciated. If we look at the coverage of land by ice with the poles in these former positions, we see that there is a net increase in land glaciated at that time. While there is a loss of ice-covered land in the southern hemisphere, there is an increase of ice-bearing land in the northern hemisphere – frozen land substituting itself for the unfrozen Arctic Ocean. So we should expect the level of the sea to have been down by the 35 metres that it was. As previously stated, this fall is totally inadequate to explain an 'ice age' in which the total ice coverage was estimated to be three times as extensive as now.

There is another puzzling feature of the 'ice age'. To achieve three times the expanse of the present ice coverage of the world would have required much more heat in the tropics in order to produce the extra water vapour necessary. As previously remarked, cold by itself will not create an ice increase, except at sea. A worldwide temperature drop has been advocated by some theorists, but this just does not fit either the facts or the requirements. Samuel Epstein and Crayton Yapp at the California Institute of Technology analysed heavy hydrogen in ancient wood: to their surprise, some twenty-two thousand years ago, winter temperatures in North America at the height of the last 'ice age' may have been higher than they are today, in areas not covered with ice. The winters were warmer, and the summers cooler.

The only conceivable condition for realising this extra ice would be if the Earth's axis were at 90° to the plane of the ecliptic: annual temperatures on the equator would be higher, leading to greater evaporation. Since there would be no seasons, there would be no summers at the poles: this would allow the ice build-up to continue without annual interruption. Even so, it's doubtful whether ice could have achieved the temperate-zone limits that the conventional ice-age theory demands. But we are just playing devil's advocate, because we know that this climatic condition did not exist. Ancient tree rings – including fossilised ones – have never indicated a total loss of seasons, even though there have been some pretty rotten ones.

# *The Earth's Magnetic Field*

**Palaeomagnetics**

In the search for pointers to world catastrophe – of the variety caused by a shifting of the poles – perhaps the most exciting discovery of this century has been that of palaeomagnetic reversals. The story begins in 1906 in a French brickyard when a geophysicist, Bernard Brunhes, discovered that, as newly cooked bricks cool, the ironrelated particles in the clay align themselves with the Earth's magnetic field. In other words these ferrous particles – in many cases magnetite – adjust their position before the brick cools sufficiently to pass through the Curie point (the temperature of 575°C, at which point the magnetisation becomes fixed and unchangeable). Thereafter the layout of the ferrous particles, behaving just like a magnetic compass, clearly points to the location of the magnetic North Pole. Subsequent research showed that not only does this configuration show polar direction, by dipping it also indicates the distance to the magnetic North Pole. For instance, a brick baked on the equator would reveal iron particles pointing horizontally towards the magnetic North Pole; one baked in the southern hemisphere would point upwards towards the North Pole (the magnetic South Pole offering more attraction to the south-seeking particles); and a baking near the magnetic North Pole would point downwards towards that pole. Brunhes' remarkable discovery became even more curious when he found that these magnetic events also occurred in lava as it cooled. Stranger still was his finding that not all historical magnetic alignments in lava pointed in the direction required by the Earth's magnetic field: some of the flows even pointed 180° in the opposite direction from where they should. Without fail, lava flows of the same period pointed in the same direction, no matter which part of the world they came from. So odd were these discoveries that many of Brunhes' contemporaries laughed him to scorn.

Some twenty years later a Japanese geophysicist, Motonori Matuyama, came to the conclusion that Brunhes had been correct. Matuyama's research over Japanese and Korean lava fields, convinced him that the magnetic field of the Earth had reversed itself on many occasions in the distant past, and at least once in the last million years. The scientific world found one reversal hard to stomach, so multiple reversals were intolerable.

Another remarkable discovery was made by Giuseppe Folgheraiter, who had actually started his research before Brunhes, in 1896. He had made studies of Greek and Etruscan vases of about the eighth century BC. These vases indicated from the aligned ferrous particles, that the vessels had been fired in the southern hemisphere. This was very odd, as they were quite obviously of European manufacture. In 1907 P.L. Mercanton from Geneva did similar tests on pots from Bavaria dating from around 1000 BC, and also on Bronze Age pottery that had been found in caves near Lake Neuchâtel. The results showed that the Earth's magnetic field was substantially as it is today: the magnetic North Pole was not too far from its present location. Clearly the pottery had been baked north of the equator. Mercanton went on to check Folgheraiter's method, and found it and its results to be as shown. Apparently the magnetic poles had reversed between the tenth and eighth centuries BC; they had then reverted to approximately their present positions some time after that.

Subsequent work in this field substantiated and enlarged on what had gone before: many previous reversals of the Earth's magnetic field had taken place. These reversals were revealed in lava, in other rocks, and in undersea sediments. But knowing about these reversals would be a lot more enlightening if only we knew how the Earth's magnetic field worked. Then, too, we need to know whether these reversals are spontaneous changes in the magnetism of the rock tested, or the outcome of change in the Earth's magnetic field itself. Laboratory experiments have identified rocks that do reverse their alignments under specific conditions. However, evidence is overwhelmingly in favour of rock-reversed magnetism being the result of a reversal of the Earth's magnetic field. So let's look more closely at the mechanics of this phenomenon.

**What produces the Earth's magnetic field?**

For want of greater knowledge, the accepted view is that the field is generated electrically within the Earth, and below the lithosphere. J.A. Jacobs writes that the origin of the Earth's magnetic field 'seems to be some sort of magnetic induction, electric currents flowing in the Earth's fluids or electrically conducting core'. He states later that 'increased earthquakes may be linked to magnetic changes'. Heirtzler, in 1970, suggests that a large earthquake could wobble the spin axis, and create a magnetic reversal; but this sounds very unlikely.

In spite of these seemingly clear pointers towards an electrical presence deep underfoot, we still do not know the source of that electricity. The pressures are so great that it is hard to envisage much movement – at a depth of 480 kilometres, there would be 220,000 atmospheres – and the temperatures are so high that any permanent magnetism, being well above the Curie point, could not exist. We are left with the belief that the magnetism must be electromagnetic.

Because of the greater fluid movement near the lithosphere, it is possible that the electrical – and hence magnetic – activity is generated largely at that level. A theory was produced by Walter Elsasser in which he attempts to explain terrestrial magnetism by convection currents. According to him, convection currents in the Earth's deep interior produce an uneven heating of the crust and thus cause thermoelectric currents to flow along the equator. A problem with this is that the convective currents are too slow to produce the expected effect. Whichever way this electric current arises, it must be remembered that it only becomes significant, magnetically, if it is related to a rotating Earth. Without that rotation no magnetic field will exist.

Adopting the accepted view, we find that the magnetic field's intensity has been steadily reducing over the last two thousand five hundred years. It is now down to 15% of its high at that time. Equally we know that the number of active volcanoes has been reducing over the same period. If volcanic activity has been reduced we can be certain that earthquake activity has lessened over a similar span. Thus movement of the Earth's interior fluids appears to be related to the intensity of the electrical activity that would generate the Earth's magnetic field.

Now, as the direction of these fluid movements must be permanent, and as the anticlockwise rotation of the Earth is equally certain, the only way the Earth's magnetic field can change is by one or the other altering. We know, for example, that in a simple dynamo the magnetic field can be reversed if the current's directional flow reverses at the same time. There is no reason why the convection-generated current flow should reverse, but if the Earth were to shift its geographical axis through 180° or thereabouts, in the way previously described, change would occur. Initially, after the axis shift, there would be no alteration in the polarity of the magnetic field; but very rapidly the solar wind – opposing the Earth's rotation in the way described in Part II and the New Theory – would slow the Earth's rotation to a standstill. Thereafter, within a comparatively short time – years not centuries – the Earth's rotation would recommence in the reverse direction. And that would mean a switch of the Earth's magnetic poles. The north geographical pole would then lie adjacent to the former south-seeking magnetic pole.

Incidentally, as stated previously, if we look at Venus, we see just such a scenario. That planet has in the not-too-distant past shifted its poles through as much as 177°: it is slowly rotating at the paltry speed of one revolution every 243 Earth days, in a retrograde mode – that is clockwise – prior to coming to rest, and then starting afresh in an anticlockwise direction. It is very likely that assessments of Venus' rotational speed in the coming years will show that the planet is rotating even more slowly, if it is not almost stationary. Its magnetic dipole moment has fallen to a reading which is less than that of any other planet (10 gauss/cm). As previously mentioned, Uranus has 'fallen over' as well, and is rotating with its axis nearly parallel to the ecliptic (its equator is at 98°). It is rotating at a speed which would be more appropriate were its axis at about 90° to the ecliptic. So its tumble, geologically, was yesterday. It now has to slow down to a standstill; regrow a new axis approximately where its equator presently lies; and recommence its rotation – this time in an anticlockwise direction.

**Relations between geographical and magnetic poles**

We have been contentedly talking as if the geographical or spin axis of the Earth is synonymous with the magnetic axis. But is this right? The magnetic North Pole moves around the geographical North Pole in a westerly direction, and is at present 11° from that pole. Do the two have an independent life? In general it is believed that the magnetic polar axis maintains a close relationship with the geographical axis: this is the view of Professor Runcorn. These two are not parallel; nor are the two magnetic poles equidistant from the two geographical poles – the southern magnetic pole is close to the Antarctic Circle, in the direction of Tasmania.

In spite of these discrepancies there is an understanding that the geographic pole influences the other by the direction of spin, if nothing else. We believe that any records showing that the magnetic poles had shifted would indicate that the geographical poles likewise had altered their location. However, it does not make sense for the polarity of the magnetic field to change without the rotational direction of the Earth changing. Even if the strength of the field went down to nil, it would still not change polarity if perchance it regained strength, without the Earth's rotational direction changing.

It's worth noting here that when a reversal occurs the strength of the new magnetic field is often many times greater than what is considered the present norm. This is because the new rotation leads to hugely increased movement of the mantle's magma. If the reversal is less or more than 180°, a new equatorial bulge has to be manufactured to replace the old one. The volcanic activity would be monumental. Naturally, this added internal flow leads to enhanced electrical activity and hence an increased magnetic field strength.

If the reversal were precisely 180°, the movement of magma would be much less – there would be no new bulge to grow – and the tendency for the Earth to almost immediately do another flip would still be there: at that moment the Earth would not have the stabilising help of a strong dipolar magnetic field.

**The reversing magnetic field**

We have mentioned that the magnetic field's intensity is but a fraction of its former high. Some 90% of that field is made up of the magnetic dipolar component. The dipolar can be likened to a bar magnet lying within the Earth along an axis that corresponds with the geographical axis. Its South Pole would be near the geographical North Pole; its North Pole near the geographical South Pole. In general, the non-dipolar field does not fluctuate in strength. However, the dipolar component or regional component of the magnetic field does undulate in strength, and this is probably related to movement within the Earth's liquid core or mantle. The main dipolar field has, as previously stated, been reducing, and according to

J.M. Harwood and S.C.R. Malin could be down to nil as early as the year 2050. We believe this implies a shift in the geographic poles, with all its terrifying consequences. It is getting frighteningly close.

According to received science, the dipolar component reverses its direction, on an average, about every 300,000 to 1,000,000 years. This reversal is very sudden by geological standards, allegedly taking about 5,000 years. The lapse between reversals is highly variable, sometimes occurring in less than 40,000 years and at other times remaining steady for 35,000,000 years, according to Professor Robert McFerron of the University of California. No regularities or patterns to these reversals have been discovered. A long interval of one polarity may be followed by a short one of the opposite polarity. We believe that reversals are far more frequent than this, and may even occur on average more often than every five thousand years, taking next to no time to complete.

During reversals, that portion of the Earth's magnetic field that reaches out into space is considerably altered. The absence of the dipolar component would mean that the solar wind could approach much closer to the Earth, meaning that dangerous radiation, not normally allowed access, would hit the surface of the Earth. The belief is that these particles might cause genetic damage in plants and animals, which could lead to the disappearance of certain species. It is our belief that the recent disappearance of many species of fauna and flora is far more likely to come from the catastrophic conditions that accompany a magnetic, and hence geographical, pole shift. What must follow the disappearing magnetic field is a general increase of world temperature by one or two degrees. So this – rather than man – is the prime cause of the present world warming. Man is offending with his release of greenhouse gas, but he is not the main culprit.

**Seafloor spreading**

Evidence for the reversals is most readily seen in the lava ejected onto the seabed from the mid-ocean ridges. These mid-ocean ridges stretch in a mainly north–south direction for some 60,000 kms through the Earth's principal oceans, marking the boundaries of the main tectonic plates. They are joined together as an almost unbroken chain of mountains; but there is a rift running down the centre of the chain, out of which pours a stream of lava. This matter flows equally to each side of the rift. It is from this mid-ocean ridge, especially apparent in the Atlantic, that seafloor spreading – the process whereby the ocean widens, and Africa and South America move apart – takes place.[8] As the rift opens more lava surges up and sets on either side of the trough. Thus we find a matching historical layout of basalt stretching away on

---

8   It must be said here that the seafloor spreading cannot be caused, as some would have it, by the pressure of the magma welling up from the mantle. Pressure of liquid basalt would never be adequate to force tectonic plates meeting on the sea bed to move apart. This continental drift, first identified by Alfred Wegener, comes as a result of the Earth's rotation. The separation is not the same at all latitudes; it could be expected to be greatest at the equator. Wegener did not believe that continental drift always continued in the same direction. He thought that the Earth's spin axis changed from time to time, which affected the direction of drift. As long as the Earth's rotation was anticlockwise (viewed from the North Pole) the drift could be expected to be westward.

each side into the distance. The flowing lava, of course, at the time of its hardening reveals the orientation of the Earth's magnetic field.

Reading the magnetic printout shown on the spreading seafloor has revealed comparatively few reversals per million years. However, some of the reversals have been for such short periods – for example, the Folgheraiter one – that they have been hard to detect. What's more, some of the changes in orientation have not been complete reversals: they have been what conventional science calls 'excursions', where the field starts to reverse, but only moves a number of degrees from the present norm. This has been described as an aborted reversal. We do not, of course, accept this interpretation. We take the view that these magnetic reversals reflect a geographical pole shift (in the Folgheraiter case, the pottery was shown to have been fired in the southern hemisphere) and that the aborted shifts reveal instances of the geographical poles relocating in regions different from the present poles. To achieve one of these 'excursions', the geographical poles would have to move through more than 90°, thus bringing the Earth's rotation to a standstill. The new rotation would be on a new axis where the poles corresponded with the 'excursion'. In the Folgheraiter case, which is well within the limits of written history, the reversal must have been in the region of 180°, because there was not the recorded catastrophe that could have been expected with the end of one equatorial bulge and the growth of a new one.

The readings of the seabed reversals cannot be taken as fully researched. As time passes more instances of reversals get discovered. Because these reversals may have been for a very short duration, prior to the normal magnetic orientation re-establishing itself, the reversal may have been missed. Then, too, magma might not have been flowing at the test location point when a reversal took place – although this is less likely, because there tends to be a considerable increase in volcanic action at the time of a geographic pole shift.

A less relevant fact should be mentioned here: the further back we go in geological time, the more unlikely different continents are to point magnetically in the same north polar direction. This seems to indicate that continental drift has moved and rotated the continents away from their present locations. Continental movement is very slow, so while this becomes significant 200 million years ago, it does not drastically alter the pattern of magnetic alignments during the Pleistocene period – the last million years.

**Recent magnetic reversals**

We are closing in on a central feature of our theme – reversals in the last few thousand years. The seafloor has revealed ancient reversals: conventional science states that the last major reversal was 700,000 years ago. However, there is healthy evidence to question this. From Lake Biwa in Japan comes data showing magnetic reversals 295,000, 180,000 and 110,000 years ago. From an earthenware hearth used by aborigines near Lake Mungo in Australia some 40,000 years ago comes more evidence of a reversal. Czechoslovakia reveals a magnetic event 30,000 years ago:

France shows one 20,000 years ago; and Gothenburg in Sweden reveals an event only 12,600 years ago.

Some of these dates are disputed. The Lake Mungo event may be more recent – perhaps 30,000 years ago; this uncertainty is due to two different dating techniques which gave different answers. Another event at Laschamps has disputed dates of 30,000 or 10,000 years ago. There is also, according to Peter Warlow:

> a single measurement from sedimentary rock data which reveals an event dated to 860 BC which corresponds with Folgheraiter's pottery data, though there has been no other confirmation of this remarkably recent date – at least not from magnetic records. All of these events occur within a period of predominantly normal polarity – the polarity of today – so they imply that either there was a large disturbance, after which the field returned to normal of its own accord, or that there was actually a pair of reversals, one to turn the field to a reversed state, and another one shortly after to turn it back to normal. In either case, the evidence of these recent reversals suggests that the time interval between events may be much shorter than the average figure of 200,000 years, and there is also evidence from earlier periods to suggest that there were similar short-term reversed episodes which the seafloor record has failed to detect.

D.H. Tarling in *Palaeomagnetism* states that 'excursions' are evident on a short timescale of 100 or 1,000 years. But the timescale is much less reliable than in the major magnetic reversals. This is partially due to these excursions being too brief to be imprinted in most oceanic sediments. The excursions are not full reversals but trips by the poles to the low latitudes, often in the opposite hemisphere. Over a dozen have been repeated in the Russian loess deposits in the last 300,000 years.

Kristjansson and Gudmunsson (1980) reported at least eight excursions in Icelandic basalt sequences in the last 120,000 years. There appears to be some evidence for a reduction in the magnetic-field strength at such times. They reasoned that excursions were abortive attempts by the magnetic field to reverse fully. This is far removed, as we have stated, from our interpretation that an excursion would indicate the direction of the new poles after the planet's rotation had come to rest. It cannot be emphasised too strongly that the only way to change the position of the magnetic pointers is by changing the direction of the globe's rotation; simply tilting the Earth will not change the magnetic alignment, because the axis of rotation and its direction are unchanged.

There is a feeling that these reports are not definitive: there may be many more undiscovered excursions during the above periods not yet identified. To talk about 'excursions' in this way is, of course, out of step with our views. An 'excursion' is not a little adventure taken gratuitously by the Earth's magnetic field: it is in fact a reflection of a major, and far more devastating, type of geographic pole shift. The full 180° pole shift may not be nearly as destructive of life and surface features, as we will shortly explain.

## *The Climate has Changed*

**What is under the seafloor?**

When we look at the underside of the seafloor, delving into the sediment lying on the bottom, we come upon a series of temperature reversals which may parallel and even complement the magnetic reversals.

The seabed, which is largely basaltic, is overlaid with ooze made up of plant and animal remains descending from the surface, and coming under the general heading of plankton. In 1872, the British Government sent HMS *Challenger* to dredge the bottoms of the world's oceans; this expedition produced a 50-volume report on its findings. Scientists were spurred on to look deeper, but this could only happen when methods of deep-sea coring were developed. It was not till 1925 that cores of one metre in length could be recovered from the top of the sediment. Walter Schott, the German palaeontologist, analysed a number of these which had been raised from the equatorial waters of the Atlantic, and identified three separate layers. Together they told a temperature story.

The top 30–40 centimetres contained an assemblage of foraminifera (limy animals whose skeletons cover much of the temperate and tropical seabeds) very different from what lay below. This top layer was identical to what is raining down from the surface at the present. The next layer contained some of the same forams, but it also had a large number of forams that normally live in cold water. One particular warm-water species – *Globorotalia Menardii* – was present in the top layer, totally absent in the second layer, and reappeared again in a third, deeper layer. Schott decided that this middle, *Menardii*-free layer corresponded with the last ice age. Layer three equated with the last interglacial, when the *Menardii* were again able to breed in a suitable temperature. It is difficult to put dates to the three core layers. No one knows whether the sediment has been raining down on the seafloor at an identical speed at all times. However, as a starting point it must be assumed that this is the case.

We, of course, do not share Schott's view. It is our belief that the three sedimentary layers point to two shifts in the location of the geographical polar axis. What's more, there is a very interesting pointer that cannot be explained by Schott's interpretation. In 1947, a Swedish crew of scientists sailed a ship called the *Albatross* around the world, and collected longer cores from every ocean. On analysis these cores showed layers of varying concentrations of calcium carbonate (lime from planktonic skeletons) at successive depths. However, the curious feature of these cores was that those from the Pacific revealed concentrations that were the opposite to those from the Atlantic. Any attempt to explain this fundamental difference by suggesting that the circulation patterns or the salt content of the two oceans are different is inadequate. No improved theory has been forthcoming.

We believe that a polar-axis shift, in which the equatorial region of the Atlantic moved north towards the frigid zone, would occasion a corresponding southward movement of the northern region of the Pacific – and the further west we examine this ocean, the more pronounced would that southward shift become. In changing its poles the Earth would have pivoted on an axis running through the equator from approximately Lake Victoria in Africa to Christmas Island in the Pacific, in this particular instance. To determine exactly where these pivot points were, we would have to identify the precise spots from which these cores came.

In 1956 David Ericson of the Lamont Geological Observatory, Columbia University, confirmed the findings of Walter Schott. He examined a vast number of low-latitude cores from the Atlantic. Together with colleagues, he pronounced that the abrupt boundary separating the two upper layers of sediment could be dated at 11,000 years ago. This date was very close to the radiocarbon date given for a sudden change of temperature on land: in other words the end of the last 'ice-age'. Ericson identified six different layers of alternating *Menardii* and *Menardii*-free sediment over the last 300,000 years – producing an averaged similar temperature span of some 50,000 years between temperature reversals. Of course, the layers were not like that; their thicknesses varied considerably.

A different approach to the temperatures revealed within the sea-bottom cores was taken by Cesare Emiliani, at the University of Chicago, in 1955. His method was to analyse the skeletal remains of forams for their oxygen-isotope content. One of these oxygen isotopes (oxygen-18) is heavier than the other (oxygen-16). Both types are present in the calcium carbonate skeletons of forams. It had been demonstrated that the amount of the heavier isotope that an animal extracted from the sea depended on the temperature of the water. In cold water the skeletons have higher contents of oxygen-18. Measuring the ratio of the two isotopes in a foram would indicate the temperature of the sea during its lifetime. Emiliani's research produced layers of different temperatures, much as had that of Ericson.

The two sets of findings had a similarity over the most recent periods of sedimentary deposition; however, over the last 300,000 years their findings departed from one another. Emiliani's found no less than fourteen abrupt changes of temperature – giving an average of 21,000 years between temperature reversals. From our standpoint this would mean fourteen occasions on which there had been pole shifts of the extreme kind – the type where a new polar location had to be found, meaning probably a move in the Earth's axis of nearer 90° than 180°. As we have said, a shift of anything approaching 180° would not have led to much change in temperature, nor would it have called for a new equatorial bulge: so the disturbance to the Earth's surface would have been minimal. Each degree of pole shift down from 180° would have caused greater temperature change in the core readings.

It is sad how extremely difficult it is to find evidence of events parallel to these temperature reversals on land. Tree rings can only provide information on climatic changes back to the years immediately before the Christian era. The bristlecone pines and sequoias that go back several thousand years do not give the evidence we are seeking. Mind you, a near 180° reversal would not necessarily leave dramatic information on the rings of a tree, since the temperature or humidity changes might not be excessive.

The geophysical and geological evidence we have amassed for the late Pleistocene period has shown pole changes/dramatic climatic changes which can be averaged to every 20,000 years (not that averaging means very much). Because we know of magnetic reversals that are not reported in the mid-ocean spreading magma, nor in the sedimentary columns retrieved from the ocean corings, we must conclude that there were pole shifts which have been very short-lived (in that the poles have flipped back to their prior positions) and others which have scarcely altered the climate. We also know that retrieving samples from the sea bed is difficult, and short reversals could easily be overlooked. Then, too, turbulence and current movement can drastically alter the way sediment is laid down. It must not be thought that a shift in the Earth's axis could be an innocuous event. However, at certain spots on the Earth's surface little might be noticed. If, for example, there was a 180° shift which would result in the two poles changing place, the pivotal points would be on the equator.

Someone living at one of these two points would scarcely notice the change, in the short term. In the first place, the Sun would not be rising in the west and setting in the east, until the solar wind had stopped the Earth's rotation and restarted it in the opposite direction, as was previously explained. Another person living some 6,000 miles further round the equator would have a very different experience. He would travel through the polar region prior to finishing at the equator again, but on the opposite side of the globe. As he would be dressed (or undressed) for the tropics, he would have a seriously uncomfortable time as he passed through one or other pole. This assumes that the axis shift would only take a day or two. It was reversals of just such rapidity that froze many of the mammoths of Northern Siberia, some 11,500 years ago, and again 3,700 years ago.

What's more, similar catastrophes hit that lengthy stretch of Russia/Siberia, freezing yet more mammoths on a number of occasions and at different locations up to 50,000 years ago (and who knows how far before that). Radiocarbon-14 dating of mammoth remains is fairly specific here. This is more evidence of the incessant fidgeting in which the Earth delights.

**The Holocene weather**

The rotating Earth exhibits a curious phenomenon called the Coriolis effect. At its equator the Earth has a speed of 1600 km/hour in an anticlockwise direction, while at its geographical poles it is stationary, and simply turning round. Anyone who has been at the equator, and who decides to move to the north or south of that line, will find himself moving faster than those people who are rotating with the earth at higher latitudes.

Consequently he will tend to move to the east. Likewise, anyone travelling from the poles towards the equator will find himself moving more slowly than the 'prevailing' land, so he will be gradually left behind, finding himself moving towards the West. This Coriolis effect is highly notional when it comes to the very slow movement that humans make, but is of some significance when it comes to aircraft navigation, or weapon firing.

The Coriolis Force is also significant when we examine the movement of winds and ocean currents. This subject of wind movement is intensely complicated, since land masses interfere with wind directions, as does height in the troposphere and stratosphere. Nevertheless, there is a tendency for winds leaving the equator towards the north to move in a clockwise direction, while those going southward move anticlockwise.

Although those winds will modify their directions when approaching high-pressure zones or depressions, nevertheless, they will have a general direction which enables us to talk about prevailing winds. To change the direction of those prevailing winds calls for something very strange to happen to the way the Earth rotates.

A study of European peat bogs by the two Scandinavians Axel Blytt and Rutger Sernander produced fascinating information on climatic change. They observed layers of pine tree stumps, some oxidised, others simply buried in the dark peat bogs. The conclusion drawn was that oxidised stumps indicated a drying out of the bogs, while non-oxidisation showed a wet climatic epoch. To create these distinctions the prevailing wind in Europe would have had to come from two different directions: that from the Atlantic would carry the moisture; the other, from Continental Europe, would be dry.

Professor H.H. Lamb in *Climate: Present, Past & Future* identifies four distinct climates in the lapse between the 'ice age' and now. The first, pointing to prevailing weather coming from the continental East, referred to as the Pre-Boreal, would have signified the end of the glaciation; the second would have started at around 6500 BP with a prevailing wind coming from the Atlantic; the third, the Sub-Boreal, again came from Continental Europe; while the fourth started at around 3000 BP, coming from the south-west, and is referred to as the Sub-Atlantic. These round-figure dates are very imprecise, but they do point to four different directions for the prevailing wind. As we have made clear, the Coriolis force would be instrumental in determining those prevailing winds. To change those wind directions, the Earth would have to change its direction of rotation. And that is exactly what we believe happened on those four occasions.

# *History Speaks of Catastrophe*

### Man writes of change

It is difficult looking for catastrophes within the time allowed by written history, or yet earlier where the historian is reporting hearsay of a prior period. Nevertheless, most racial histories talk of such disasters. The reporting is often flowery and poetic, making fantasy and fact hard to separate.

Mythology of the Chinese, the Indians, the Old Testament, the North American Indians and the Inuit talks of floods and tempests that have led to racial elimination. The biblical flood is an instance of the problem. Certainly it was a disaster; but to account for a flood to the required depth – inundating mountains – after forty days and nights of torrential rain is stretching the imagination. Other hurricane characteristics are needed, and the sea must have been embroiled. So where is it possible to look for facts that are separable from myths. Then, too, will these often questionable facts detract from the more scientific side of this book?

We think there's room for both facets of approach. If the anecdotes of written history can be substantiated by palaeomagnetic readings, geological evidence or other geophysical findings, the case hardens.

There are two important pieces of reporting that seem to comply with this approach. Both Plato and Herodotus produce written accounts of much earlier events that imply devastation and change in the Earth's rotation. Herodotus in Book II of his *Histories* quotes from meetings that he had with Egyptian priests in the second half of the fifth century BC. They told him that the period following their first king covered three hundred and forty-one generations. Herodotus calculated this to be a span of some 11,000 years (a trifle generous, perhaps). The priests stated that 'four times in that period the Sun rose contrary to his wont: twice he rose where he now sets, and twice he set where he now rises'.

Plato in his *Timaeus* – writing 150 years after the event he describes – records how Solon, the wise ruler of Athens who was a friend of Socrates' great-grandfather, on a visit to Egypt questioned the priest-historians of that country.

It's worth quoting Professor B. Jowett's translation verbatim:

> Thereupon one of the priests who was of a great age said: O Solon, Solon, you Hellenes are but children . . . Solon in return asked him what he meant. I mean to say, he replied, that in mind you are all young; there is no old opinion handed down among you by ancient tradition; nor any science which is heavy with age.

> And I will tell you why. There have been and will be again, many destructions of mankind arising out of many causes: the greatest have been brought about by the agencies of fire[9] and water and other lesser ones by innumerable other causes . . . At such times [of great conflagration] those who live upon the mountains in dry and lofty places are more liable to destruction than those who dwell by rivers or on the seashore. And from this calamity the Nile, who is our never-failing saviour, saves and delivers us.

---

9   Destruction by fire would result from the rotation of the Earth stopping with a specific region facing the Sun. Days and nights of unbroken sunlight beaming down on one place would create devastating temperatures, and their attendant fires.

When on the other hand the gods purge the Earth with a deluge of water, among you the survivors are herdsmen and shepherds who dwell on the mountains, whereas those who, like you, live in cities are carried by the rivers into the sea.

But in this country, neither then nor at any other time does the water come down from above on the fields, having always a tendency to come up from below, for which reason the things preserved here are said to be the oldest . . . And whatever happened either in your country or ours, or in any other region of which we are informed – any action which is noble or great or in any other way remarkable – has all been written down by us of old, and is preserved in our temples.

Whereas you and other nations are just beginning to be provided with letters and other requisites of civilised life, when at the usual period, the stream from heaven descends like a pestilence, and leaves only those of you who are destitute of letters and education; and thus you have to begin all over again as children, and know nothing of what happened in ancient times, either among us or among yourselves. As for these genealogies of yours which you have recounted to us, Solon, they are no better than the tales of children; for you in the first place remember one deluge, and there were many of them.

This passage of Plato's does not link catastrophe with pole shift. Moreover, it does not produce a trigger mechanism for destruction by fire and water. However, there is a passage in *Statesman* – in which it is a trifle harder to separate myth and true reportage – where in an earlier age, 'the Sun and the stars rose in the west, and set in the east, and that the god reversed their motion, and gave them that which they have at present'. Later he says, 'There is at that time destruction of animals in general and only a small part of the human race survives'. The changing movement of the Sun was referred to by many Greek authors before and after Plato. In addition, Egyptian papyri support each other on the subject: the Harris papyrus reports a cosmic upheaval when 'the south becomes north, and the Earth turns over'. The Ipuwer papyrus states that 'the Earth turned upside down', and bewails the terrible devastation caused by the work of nature. Similarly, the Ermitage papyrus in St Petersburg refers to a catastrophe which turned 'the land upside down'.

**Some 3,500 years ago**

It is highly likely that the penultimate reversal of Herodotus corresponds with the calamities reported at the end of the Egyptian Middle Kingdom, and during the Hyksos period of rule. The Talmud tells of great disturbances in the solar movement at the time of the Exodus, which was thought to be the period we are talking about. The Midrashim also speaks of a disruption in the solar movement on the day of the passage (when the Israelites passed through the 'dry' Red Sea).

These accounts of solar change also refer to the clouding over of the Sun at the time of the reversals. In quoted cases there was a thick cloud cover, which lasted up to twentyfive years. Faint light or no light at all seems to have prevailed during this part of the Hyksos period: the Midrashim states that during the wandering in the desert the Israelites did not see the Sun at all because of the clouds; and they were unable to orientate themselves on their march. The Egyptian papyrus known as Anastasi IV contains a complaint about gloom and the absence of solar light; it also says, 'The winter is come instead of summer, the months are reversed and the hours are disordered.' Plutarch gives the following description of a change to the seasons: 'The thickened air concealed the heaven from view . . . and the Sun was not fixed to an unwandering and certain course, so as to distinguish orient and occident, nor did he bring back the seasons in order.'

We should remember that one of the best-preserved frozen mammoths dates back to approximately (plus or minus) 1700 BC, which suggests that the North Pole moved close to its present position, not far from the Siberian coast (having been to the South Pole first, as we explained). It is also possible that the polar movements were the opposite of this, if that particular pole reversal was not succeeded by two later changes.

Other traditions also report the strange behaviour of the Sun. In China during the reign of the Emperor Yao – possibly earlier than the Hyksos period – a world catastrophe brought disaster upon eastern Asia: for ten days the Sun did not set. A non-setting Sun is what one might expect on the opposite side of the globe from where there were reports of prolonged stretches of darkness. Clearly, the rotation of the Earth would have come to a halt at that time, before starting again in the opposite direction, impelled by the solar wind in keeping with the New Theory. According to Bellamy, the Chinese stated that there was a time when 'the signs of the Zodiac have the strange peculiarity of proceeding in a retrograde direction, that is, against the course of the Sun'.

The hieroglyphs of the Mexicans deserve mention. They describe four movements of the Sun; these refer to four prehistoric suns, or world ages, with shifting cardinal points.

The change from east to west of the Sun's rotation would be combined with replacing north with south. As we have said, without turning the Earth upside down there would be no method of changing the direction of the rotation, since the solar wind would not have changed the way it impacted the surface of the Earth. The Eskimos of Greenland told missionaries that in an ancient time the Earth turned over, and the people who lived then became antipodeans, according to Ulrik in the Scandinavian myth of Ragnarok. There is also a quaint little anecdote, cited by I. Donnelly, from the Aztecs:

> There had been no sun in existence for many years. The chiefs began to peer through the gloom in all directions for the expected light, and to make bets as to which part of the heavens

the Sun should first appear in. Some said 'here', some 'there', but when the Sun rose, they were all proved wrong, for not one of them had fixed upon the east.

These accounts, although written in a rather woolly way, do seem to have a common theme running through them. There is enough evidence from enough sources for us to believe that these writers were reporting actual worldwide changes. So it begins to look as if the major catastrophes of the world's past, in many instances, can be linked to shifts in the Earth's axis of rotation.

Two or three centuries after the fall of the Egyptian Middle Kingdom, Queen Hatshepsut came to the throne. Her vizier, Senenmet, gives an insight into changes that had befallen the Earth. His tomb had two charts of the sky painted on its ceiling; one chart is of the northern hemisphere; the other shows the southern hemisphere, with the signs of the zodiac and other constellations in a reversed orientation.

The trouble is, no one knows which hemisphere applied to the time of Senenmet's life. Did the crisis at the time of the Hyksos turn the Earth's axis to reveal the northern hemisphere to an Egyptian then living, or did Egypt move into the southern hemisphere at that moment? On balance we believe the Earth had been reversed prior to the Hyksos changes, and was then returned to a position where the North Pole was not far from its present position. The reason for this belief is that some five and seven centuries later – approximately 1,000 and 800 BC – the Earth did two more polar reversals, according to the evidence of Mercanton and Folgheraiter, ending up at its present position. So this would suggest that the Hyksos reversal brought the Earth back to its present norm.

It's worth quoting A. Pogo on this subject: 'A characteristic feature of the Senenmet ceiling is the astronomically objectionable orientation of the southern panel.' The centre of this panel is occupied by the Orion–Sirius group, in which Orion appears west of Sirius instead of east. 'The orientation of the southern panel is such that the person in the tomb looking at it has to lift his head and face north, not south . . . With the reversed orientation of the south panel, Orion, the most conspicuous constellation of the southern sky, appeared to be moving eastward, i.e., in the wrong direction.'

Modern astronomy does not for one moment admit or consider the possibility that at some time in the past east and west were interchanged, as were north and south. In consequence, the chart of the southern hemisphere viewed from Egypt makes no sense at all. The fact that the northern hemisphere shows that the Pole Star is in Ursa Major is puzzling too. If one makes allowance for the movement of the axis by precession – where one circuit of precession takes some 26,000 years – the Pole Star some 4,000 years ago would be near the present Draco, certainly not in Ursa Major.

## 1000 BC to 500 BC

According to Seneca, the Roman philosopher, statesman, orator and tragedian who virtually ruled Rome in the middle of the first century AD, the Great Bear had formerly been the polar constellation. After a cosmic upheaval shifted the sky, a star of the Little Bear became the Pole Star. We believe, whatever contemporary received science says, that here is an authoritative person who knows something that the present-day researchers do not.

Hindu astronomical tablets compiled by the Brahmans after 1,000 BC show a deviation from the expected position of the stars at that time, according to J. Bentley. Modern scholars were puzzled by this, because even allowing for the precession of the equinoxes the methods employed by the meticulous Hindus would not explain these differences.

In the Jaiminiya Upanishad it is written that the centre of the sky, or point round which the firmament revolves, is in the Great Bear. This, of course, supports Seneca. Hindu sources go on to say that the displacement of the Pole Star was between 500 and 900 miles measured on the Earth's surface, which would have been just sufficient to move that 'pole position' from the Great Bear to somewhere in the region of Draco, which is where we would have expected it to be prior to the precession which has occurred since then.

If one wishes to determine the seasons at any given latitude, one measures the length of shadow cast by the

Sun at the winter and summer solstices. The winter shadow, for instance, is longer than the summer one; and as one goes further north that winter shadow increases. By measuring these shadows, the latitude of the measurer can be determined.

Ancient astronomical tables from Babylon during the eighth century BC provide exact data, according to which the longest day was equal to 14 hours 24 minutes, whereas a summer solstice day measured nowadays at the ruins of Babylon would give a duration of 14 hours 11 minutes. This is very informative: it shows that Babylon used to be sited 2.5° further north than it now is. This is not a great distance – some 175 miles. However, if one goes back to the Mercanton 78 and Folgheraiter discovery, it looks as if the Earth had turned turtle by this time and that Babylon was established in a new location, in the southern hemisphere. It would have been shortly after these records were produced that the Earth again shifted its poles and returned Babylon to the northern hemisphere, albeit 175 miles nearer the equator. It's interesting that Johannes Kepler knew of these ancient records, and pointed out that Babylon had not always been at the same latitude.

It is worth going through an exercise which makes the assumption that the celestial movement described by the Hindus and the movement of Babylon were part of the same pole shift. The North Pole appears to have moved more than three times as far as Babylon. If we look for pivotal points on the equator that would allow those two movements, it transpires that the centre of Zaire, or the ocean due south of

Sri Lanka become strong contenders. (This also assumes that Babylon, to achieve a southward movement of 175 miles, would take a direct north–south path.) All this implies that prior to the last pole shift, in the eighth century BC, the North Pole would have been somewhere near the New Siberian Islands in the East Siberian Sea, or, if the Earth had pivoted in the opposite direction, near the centre of the northern half of Greenland.

**Earlier calendars**

In the New Theory of Part II, we touched on the way the solar wind activates the rotation of a planet. This thrust by the solar wind brushes the surface of a planet tangentially on both sides of that planet, which would seem to cancel out any rotation. Nevertheless there is a net thrust on the 'anticlockwise side' causing the planet to rotate in that direction. As previously pointed out, this thrust varies enormously in its efficiency. If, in the case of the Earth, the equator was parallel to the ecliptic – the line in which the solar wind is flowing towards the planets – and not tipped by 23.5°, the solar wind would be at its most efficient for creating rotation.

We could then expect the number of earthly rotations per year to be much in excess of the 365 days and nights we are used to. The year would, of course, be of the same duration, but its days would be more rapid and last marginally less time. As it is, the inefficient angle of 23.5° keeps us at the present lowly figure. If we could point to a time when either this angle – and of course we infer from this the angle of the geographical poles – was different, or the rotation period was greater or less than 365 times per year, we would also find a different calendar for this period, and this would substantially support the argument for a pole shift. We must emphasise here that this is not calling for a change of location of the poles on the Earth's surface, but a change in the sidereal directions in which those poles are pointing.

We believe that it is possible to identify a time when both polar angle and location changed. Folgheraiter and Mercanton found evidence of two magnetic pole changes, somewhere around 1,000 BC and then around 800 BC. We believe that magnetic pole changes, as we have argued, mean the geographic poles have changed at the same time. That disasters, involving absence or excessive presence of the Sun, occurred at this time seems obvious from the Talmud and the Bible. In the reign of King Uzziah there was a devastating earthquake which killed thousands, not just in Israel and Judah, but in many other parts of the world. Subsequently, the calendar was changed in that country, in the year 747 BC, because for some time it had been known that there were no longer 360 days in the year, but just over 365 days.

**The 360-day year**

It is very hard for received science to understand how the previous calendar, almost worldwide, had a year of 360 days. That the year could actually have lasted 360 days clashes horribly with uniformitarian thinking. Yet how could astronomers as informed as the early Babylonian-Assyrians have got things so wrong? They were not alone: the texts of the Indian Vedas refer to a year of 360 days, as do the later Brahmans. They do not refer to the missing 5.25 days, nor do they mention an intercalary period where missing days are slipped into the calendar, as we do nowadays every leap year.

The Persian year also had 360 days divided into 12 months of thirty days, as had the Indian year. Since neither mention intercalary adjustments we must assume that 360 days was a very accurate measurement of the year. Obviously it cannot be precise, but it was good enough to have been accurate for several hundred years. No calendar can be totally accurate, and, as we know, our present one would have slipped ahead of the astronomical years if over the last 400 years three of the centennial leap years had not been removed. The months of thirty days are more arbitrary and did not reflect the orbital time of the moon. Presentday efforts to attribute an accurate duration to a month are no better: a synodic month of 29.53 days would be fearfully unhelpful.

The Babylonians, likewise, had a year of 360 days which they extended to the circle apparently made by the Sun, which traversed 360° in the year, or one degree per day.

The Assyrians followed suit. The Egyptians had a year of 360 days which had been used since the upheavals in the time of the Hyksos, after the end of the Middle Kingdom.

What the length of year was before this date, we do not know. The Egyptians appear to have properly adjusted their calendar to something akin to the Julian Calendar in the year 238 BC, which was long after the solar year had changed.

A decree was drawn up at Canopus, and inscribed on a tablet, in which an extra day was added to the year every fourth year (the extra five days were added later). This tablet is more famous for the three languages in which it is inscribed: Greek, Demotic and Cuneiform – by many it is regarded as second only to the Rosetta Stone in importance.

It seems that the Egyptians were very reluctant to give up their 360-day year: until this decree, they continued to work with twelve months of thirty days, adding the extra 5 days as an unmentionable, unpropitious period at the end of each twelve months.

Across the Atlantic the Mayas had a year of 360 days, prior to adding five extra days and a sixth every four years. In South America, according to Markham, the same story prevailed – a 360-day year and later 5.25 days added. On the other side of the Pacific the Chinese had a 360-day year which changed to 365.25 days. Throughout the world the same story prevails. Some time around the eighth century the Earth changed its rate of rotation, speeding up to 365.25 rotations in a year.

We believe that the evidence is overwhelming that the Earth changed the angle of inclination of its poles at that time. Whether the poles changed places – as we have said, this looks likely – or not is not relevant at this juncture. What we are saying is that the polar angle to the ecliptic became more obtuse. It changed from something less to nearer its present 67.5° angle, thereby making the tangential action of the solar wind more effective. This of course stepped up the rate of rotation.

Conventional scientists who examine this matter either arrogantly assume that the earlier astronomers and mathematicians couldn't count; or, if they grant them that concession, they believe the Earth was on a shorter orbit. They never consider that the diurnal duration could have been longer, i.e. the rotational speed slower. It is always assumed that the Earth's speed of rotation is invariable. Yet with the polar star in Ursa Major the angle of tilt would have been greater, and – in step with the New Theory – the speed of rotation would have been less. Should we not be wondering why our globe is now divided into 360° – something inherited from much earlier map makers to correspond with the days in the year. After all, that number is not metrically convenient, and doesn't seem to have any relevance.

# Have we Made a Case for Catastrophe?

### The distant past clearly indicates calamities

Looking back into the dim reaches of the Earth's past, we see abrupt cut-off points at the end of each geological period. Many varieties of fauna and flora that have flourished in one period do not exist in the next (that is, of course, how the classifications are made). Sometimes these changes appear, many millions of years in the past, to be very sudden – although 'sudden' is hard to assess when looking at rock strata. Do we mean changes that occur over thousands (and even millions) of years, or do we mean abrupt changes – even overnight? Certainly, the interval between the end of the Cretaceous and the start of the Palaeocene appears to be very short. This was a time when there occurred the single largest destruction of animal life known to science.

It seemed that 75% of the world's creatures – on land and in the sea – were wiped out in one sweep. Because there is evidence of a magnetic reversal at this time, we should examine the event. That this catastrophe was not brought about by the methods we have been arguing for seems likely. Nevertheless, a disaster which was followed by a switch in the Earth's axis is highly relevant to our thesis.

**The dinosaurs**

The Cretaceous period ended some 65 million years ago. Prior to that moment the Earth had an animal population that included many varieties of dinosaur. The following period – the Tertiary – reveals a total absence of these giant reptiles. What happened to them?

The town of Gubbio is famed for its ceramics; the clay used for this pottery and brick-making dates to the end of the Cretaceous and the beginning of the Tertiary period. Chemically, it is rather unusual. Scientists examining the Gubbio clay found it contained significantly large amounts of the precious metals osmium and iridium, which belong to the platinum group of noble metals.

These metals are dense, heavy and rare at the surface of the Earth. However, they are not so rare in other parts of the solar system: they will be found at the centre of the Earth, but, more importantly for this subject, they will be found in asteroids normally resident in the asteroid belt, and orbiting between Mars and Jupiter. If one of these should hit the Earth and disintegrate, the heavy elements from its interior would be found on the surface of the Earth. And that is a likely explanation of what happened at the end of the Cretaceous. Samples of the Gubbio clay were analysed at the Lawrence Radiation Laboratory of California University at Berkeley. They were found to have thirty times as much iridium as in the clays above and below the 65-million-year level.

Scientists needed to find more of this boundary clay in other parts of the world to establish the arrival of the extraterrestrial visitor. South of Copenhagen and in the North Island of New Zealand similar boundary clay was found showing iridium and osmium in volumes far above that of the adjacent strata. Before long the Berkeley team had satisfied themselves that round the world there were places where this enriched clay confirmed the Cretaceous boundary, not only on land but in the oceans as well. Was it just coincidence that so many creatures had died at the same time as this arrival? The verdict was 'no'; they had died as a result of the asteroid.

The death of the dinosaurs would not have been due to the impact; the dust raised on collision would have blocked out sunlight, terminating photosynthesis for several years. So, first the herbivores would have died, then, once their carcasses had been eaten, the carnivores would have gone.

That there should have been a magnetic reversal at the same time as the impact is not surprising. With little doubt the asteroid would have caused the Earth to shift geographical poles. This only goes to show how fragile is the Earth's balance; life would have been made that much worse by the horrors that went with geographical-pole reversal. In this instance the shift would have come about in a way that was different from what we believe is the norm. Asteroids may hit the Earth from time to time, but they do not cause havoc as frequently as pole shifts do by the other methods we have described.

It is unlikely that asteroids as destructive as that one made a habit of impacting the Earth. We believe that shifting of the geographical axis was something that happened – if not with regularity – on a constantly repeating basis. So the ends of most geological epochs signal just such events. That is not to say that the ends of these time spans were the only catastrophic times the Earth experienced. The pointers identified in the Recent epoch should give guidance to the often-repeated violence of all eras. In the latter part of the Pleistocene, there were abrupt sea-temperature changes that occurred on average every 21,000 years (according to Cesare Emiliani). The acknowledged magnetic reversals of the last 50,000 years come closer to averaging intervals of 10,000 years. It is when we look to historical times that we find reports of the Sun rising in a different place on four instances within the 10,000 years (marginally adjusting Herodotus' calculation) prior to 600 BC.

Is it unreasonable, therefore, to expect some sort of catastrophe – it would obviously vary in amplitude from one time to another, also from one place on Earth to another – every 2,500 years. We think not. And that means that the next doomsday event is now due, *or even overdue*.

**The doomsday process**

Mention has been made of the various options open to the geographical poles when they decide to move. We believe it is impossible for the Earth to simply shift its two poles to new quarters.[10] It seems much more likely that a full reversal of 180°, or thereabouts, takes place: this enables the solar wind to bring rotation to a halt. Thereafter, the new rotation recommences with the poles initially at right angles to the ecliptic.

Three thousand seven hundred years ago, dwarf mammoths, which had survived the disaster that overtook their ancestors at the end of the 'ice age', perished. On Wrangel Island off the Siberian coast, some 2,500 mammoths are known to have frozen – and some have been found in a well-preserved state. This slaughter seems to coincide with the Egyptian records at the time of the Hyksos. It was apparent that catastrophe in the Middle East was not on a major scale at that time; but in other parts of the world it could have been. An axis shift where the Earth pivoted on equatorial points in the Pacific south of Guatemala and in the Indian Ocean

---

10 It can, of course, alter the tilt quite easily, but with an increased tilt this simply produces a much greater swing between winter and summer temperatures: summers became far hotter at higher latitudes and winters far colder; likewise, with a reduced tilt the seasonal swings become less.

south of Sri Lanka would have led to much movement – and hence catastrophe – in other parts of the world. On such a pivot Wrangel Island would have passed through the North Pole, which would have found the necessary temperature to freeze the mammoths in the middle of a southern summer.

Anything short of a shift of 180° means the Earth, as it starts to rotate, will have to reshape its crust. Firstly, the equatorial bulge in its old form has to shrink, and as the rotational speed increases the new equatorial bulge has to form. Then, too, the oblate features of the Earth have to move: the old flattened polar regions fill out, and the new polar regions become flattened. All this means huge movements of magma below the lithosphere. The centrifugal flow of this magma could take many centuries before the substantial nooks and crannies have been filled with new material. Rather like clothes in a spin dryer, this moving magma gets pressed against the inside of the crust, which leads to terrifying volcanic activity and devastating earthquakes – far in excess of anything seen today.

It was pointed out in an earlier passage that the volcanic activity some 2,000 years ago far exceeded the activity of today. But then that activity was far closer in time to the last pole shift. Given a few centuries, the heaving of the crust lessens, and the number of active volcanoes reduces: as we know, the count is now down to some 1,600 active/dormant, of which 500 are seriously active; whereas 2,000 years ago that figure was many times greater, as we see from the thousands of defunct volcanoes that litter the Earth.

A change of poles with its attendant movement of magma within the upper mantle, and to a lesser extent in the lower mantle (the centrifugal effect is less as the core is approached), means there is far more overall internal activity; and this means greater magnetic induction. That is why the Earth's magnetic-field readings increase dramatically at a time of magnetic reversal, which equates with a geographical polar reversal. As time passes, the magnetic readings go down, in step with a reduction in magmatic movement. It has been pointed out that the Earth's dipolar magnetic field readings are reducing; if this rate of decline continues at its present pace, the field could be down to a negligible figure within 50 years, in the view of some scientists.

**Events surrounding a pole shift**

Assuming we have made a case for polar change, which is irregular but averaging every 2,500/5,000 years, let's examine the sequence of events. The axis moves, say 150°, and this means the solar wind has to bring the planetary rotation to a standstill. The first thing that happens thereafter is that the ice and snow in the Arctic/Antarctic melt: this occurs much more rapidly than the replacement ice builds up at the new poles. That confirms the findings of George Kukla, whose researching into Pleistocene climate fluctuations in the former Czechoslovakia pointed to a slow build-up of cold evidence and weather, but an abrupt change from cold to warm weather.

The loss of polar ice means that there is a reduction in albedo in the polar circles; this makes it easier for the solar wind to 'gain a purchase' on the Earth's surface, and initially to stop rotation prior to restarting it in an anticlockwise direction. Internally, magma is moving, and the new, stronger dipolar field starts to develop as the rotation begins. This means that slowly the magnetic field builds up a fence against the incoming solar wind, which ultimately limits the speed of the Earth's rotation. The protection from solar radiation afforded by the magnetic field before the poles have shifted has been poor; consequently, global warming would, as now, have been apparent.[11]

It is highly likely that the present global warming is only partly caused by man. Increased temperature is much more likely due to the loss of protection from the Earth's magnetic field. As this lessens in strength, it permits increased solar radiation to arrive at the Earth's surface. On its way in, this radiation impacts higher levels of the Earth's covering, such as the ozone layer, which is damaged by it.

The cycle continues. The polar ice, which, until recently, has been increasing over the years since the last pole shift, adds to the Earth's albedo. This means that gradually the Earth's rotation starts to show wobbles and nutations: at the start of the new rotation several thousand years before, that rotation had been wobble-free. By this time, too, the internal movement of the magma has slowed down — the new equatorial bulge having become

---

[11] This, of course, is not related to the warming created by the increased presence of greenhouse gases such as water vapour, carbon dioxide, methane and others, largely released by man. After all, man's annual output into the atmosphere of carbon dioxide is only 6,000 million tons, which is little more than that released by one large volcanic explosion.

firmly established – so the magnetic field has fallen to a low.

Once again, the incoming solar radiation starts to cause global warming. In our present situation, with a land mass at the South Pole and frozen sea at the North Pole, this global warming is having a greater effect on the northern sea ice than on the Antarctic ice. In the near future this will mean that much of the sea ice will disappear before the southern ice; the difference in albedo at the two poles will be accentuated. Obviously, this difference will have a noticeable effect on the way the solar wind/radiation spins the Earth – wobbles will be compounded. The picture will be showing all the ingredients for another pole shift.

**What is the trigger mechanism of a pole shift?**

There are two variable ingredients that cause a pole shift; both have been touched on above. The first is the Earth's albedo and the second, the reducing magnetic field; of the two, the former is the more important. The Earth's coverage of ice is on the increase. This statement has been challenged, and scientists have been at pains to say that they cannot tell whether there is a net increase, a standstill, or a net loss of this ice. Because 95% of the world's non-marine ice is locked in two giant sheets – Antarctica and Greenland – it should have been easy to measure the net changes in volume. But it's not. In 1967, the accumulation of ice showed an increase of 1,900 cubic kilometres. The ablation in that year – through melting, evaporation, carving of ice sheets and glaciers, carving of ice shelves, and bottom melting – added up to 1,660 cubic kilometres. There was a net gain of 240 cubic kilometres. But these figures were disputed. Now that we have reached the twenty-first century, there is little improvement in our ability to say whether there is a net increase or decrease in the world on-land ice.

The Antarctic peninsula has been losing ice, but elsewhere there have been gains. If there is anthropogenic global warming it may in fact be increasing the amount of onland ice. Yet it is possible that the polar build-up of ice has peaked. Needless to say, the last 2,500 years has seen steady growth. In addition to the on-land ice we must not ignore the albedo contribution of the Arctic Ocean's ice. Then, too, as has been stated, the reduced magnetic field is letting more solar wind reach the Earth's surface, which, while adding to the global warming, is also compounding the instability problem.

The solar wind hits the Earth, as we have said, on an alternating basis. The vast body of this wind impacts the tropics. However, in a northern summer the wind is brushing the northern hemisphere, in preference to the southern, and in winter the converse. We might think this balances the rotation-causing effect of the solar wind, even if done in an alternating and wobblesome way. This is not the case. An uneven solar radiation, compounding its unevenness by brushing a surface that has high albedo, raises the chance of making the Earth tumble.

The second reason for a change in the Earth's geographical axis is the reducing dipolar magnetic field.

As the magnetic strength falls away, the protective canopy of the field on the sunward side of the Earth becomes less effective. This allows the solar radiation in greater volume to impact the Earth. In its turn, this increases the chance of the solar wind delivering an uneven thrust to the Earth's surface.

Wobbles are compounded. Someone might point out that a reduction in the magnetic field should lead to a reduction in the Earth's orbital speed. Theoretically, this could be the case. In fact the Earth's magnetotail is so lengthy that the solar wind will always have something to impact. This magnetotail will still have a length of several million kilometres, but it will not have quite its former (east–west) bulk.

If the Earth turns through a full 180°, the likelihood of another reversal occurs. Firstly, the polar ice sheets do not melt; secondly, the equatorial bulge does not alter; and thirdly, the magnetic field – experiencing little movement in the magma beneath the lithosphere – remains weak. So here we have all the ingredients for another polar reversal. It follows that a movement of less than 180° creates a crescendoing manifestation of violence at the Earth's surface, which becomes more pronounced if the pole shift approaches 90°. At this point, of course, the polar axis lies parallel to the ecliptic.

**The Earth cleanses itself**

However the axis shift occurs, damage is done to those who occupy the surface of the Earth. Any reversal of the poles – a shift of 90° or more – calls for a cessation of rotation, followed by a start-up in the opposite direction. This demands a period of prolonged night or prolonged day. The former is uncomfortable, cold, and bad for farming. The latter is far too hot and could lead to dangerous conflagration. There is also a twilight zone between the two extremes; obviously that is not very satisfactory, either. In addition, there may be severe clouding caused by the dust raised in the act of pole shift. Volcanic action will have compounded this obscurity. Then, too, we must remember that unless the new poles are tipped several degrees less than perpendicular to the ecliptic, there will be no seasons.

Maybe we don't need them. All the same, it seems traditional that the world of plants prefers it that way. It must be said that the solar wind would tend to start rotation with the poles perpendicular to the ecliptic; then almost immediately a tilt would occur. There is negligible evidence for a seasonless world.

These oft-recurring pole shifts, as we have stated, can do the immense damage which was described earlier on. Equally, they may have little effect on specific spots on Earth.

Generally they must be seen as a cleansing operation. After a prolonged period in one position the surface of the Earth will be running short of carbon dioxide vital for the photosynthetic life of plants. A polar shift, with its concomitant volcanic activity, leads to a substantial release of this gas. Then, too, the storms accompanying these movements will have eliminated much animal and human life. As horrifying as this is, the land appears to need a period of rest and regeneration, prior to regrowing its inhabitants.

**History is lost**

These catastrophes mean a wipe-out of history. The pole shifts of some 2,700 and 3,500 years ago did not eliminate too many people. However, it seems evident that what happened at the end of the so-called last ice age not only removed mammoths, but most of the other animals and humans too. This sort of destruction also means the end of any technological development. The death of a civilisation compels its successors to begin again with a stone-age level of existence. Is it too speculative to suggest that others in the distant past have reached our level of knowledge? After all, from the Neolithic level to our present is only some 7,000 years. It is interesting to contemplate the technological levels that might have been achieved prior to the crisis that changed the Earth some 11,500 years ago. It is also interesting to speculate where that development might have been: could the north-east coast of Siberia be a good starting place?

# *Summary*

It is hoped that a case has been made for saying that the Earth shifts its poles constantly. Since the end of the last 'ice age' (which never existed) the poles have changed at approximately 11,500 years BP, 3,700 years BP (plus or minus 200 years), then 2,900 years BP, followed quickly by another reversal some 2,700 years BP.

Magnetic-pole reversals are evident at each of these dates. An abrupt change of marine-water temperature supports the earliest date. That there have been dramatic changes in the level of the sea can only be explained in terms of ice freezing and melting; and the fact that ice melts when it ceases to be in the polar region must be a *sequitur*.

We have reasoned that there is no way for the magnetic field to reverse unless either there is a change in the direction of flow within the Earth's magma – which would not be possible in the timescale available – or there is a change in the direction of the Earth's rotation. The latter is feasible. So magnetic poles only change when geographical poles change.

Lastly, history has pointed to change: the Sun appears to have risen from different directions on four occasions within the last 11,500 years. The calendar has changed its year span at a time that corresponds with the last pole shift.

The case rests.

# *Afterthought*

### People prior to the end of the 'ice age'

It's a strange old world we live in. If we have made a successful case for pole shift, at irregular time intervals, we are also looking at a repeated build-up of civilisation – technological development, if not enhanced humanitarian behaviour – followed by a ghastly destruction of humanity, and the loss of that technology.

It seems that there was death and destruction at the time of the last three pole shifts – two between 2,500 BP and 3,000 BP, and a third around 3,500 BP. This was not on the scale of the devastation of 11,500 BP. However, we have more forensic evidence of destruction between 3,500 BP and 4,000 BP: the death and freezing of the dwarf mammoths on Wrangel Island is very indicative of a pole shift through some 180°, with its dire consequences. Equally, the increasing desertification of the Sahara, the Nevada Desert, and the Taklimakan meant the end of human habitation in those erstwhile rich and cultivated regions. By contrast, the destruction of human and animal life at the end of the socalled ice age was very apparent, and is generally accepted by received science. So what existed in that perfectly habitable climate, which was no colder than that of today, immediately prior to the catastrophe which signalled the end of the 'ice age'?

We know that there is much evidence in many parts of the world, from that time, of the existence of Upper Palaeolithic people, whose artifacts did not extend beyond stone tools and weapons. There is less evidence of peoples who had developed higher technologies. But there is some such evidence. And it is not unreasonable to believe that the world of that time could have had both savage and civilised members. After all, in the eighteenth and nineteenth centuries, 'civilisation' in western Europe was highly advanced, while in central and southern Africa a stone-age world existed; equally, in Australia the aboriginals showed a comparable lack of technology.

Why should so little evidence of earlier habitation be available? Put simply, a cave lends itself to being a readily accessible shelter to someone incapable of building a dwelling for himself. Caves last longer than man-made houses, and their contents often survive for longer periods. However, they are not comfortable places to live permanently – they can be dank and dark. So Cro-Magnon man, for example, living within the last 35,000 years, would have built his own home; that house would neither have survived nor left any traces after the catastrophe that struck the Earth some 11,500 years ago. Land would have subsided, while other lands would have risen. Seas would have surged over dry land: that man tends to perch himself near river, lake or sea meant that water would have added to the destruction accompanying unbelievable wind movement and earthquakes.

Now, Cro-Magnon is the name given to a number of skeletons found in a rock shelter at Cro-Magnon, near Les Eyzies-de-Tayac in the French Dordogne. On the limited strength of this discovery in 1868 a general overview of prehistoric man of this period was established. The Cro-Magnon remains were asserted to be representative of a Cro-Magnon race, and therefore typical of human beings populating the Earth – possibly in small numbers – at that time. It was discoveries from this period that led to its human beings being described as stone-age or cavemen. We know that even today, looking round the world (with its far-reaching interconnectedness), that this theory does not hold up. Cro-Magnon could have been one of many races with varying levels of technological development. Then, too, it must be remembered that during the Upper

Palaeolithic Britain was under an Arctic ice sheet – if our arguments are correct – and Cro-Magnon man would have been just south of the Arctic; so the climate would scarcely have been temperate. That is not a good latitude for developing a high technological civilisation, where most of man's energy would be directed towards keeping warm.

All the same, those people did manage to develop a high level of artistic culture, as we know from the cave paintings of southern France and northern Spain. These, together with those of the central Sahara, can be dated as far back as 25,000 BP. That said, if we were to look towards the temperate zone, there might be indications of greater technological development. For such development to survive the buffeting of the winds evident at the end of the 'ice age', it would have to be something very substantial; the suggestion is that megalithic structures are worth examining.

**Megalithic evidence**

There is little to be said about the menhirs and dolmen that can be seen all over western France, Ireland, Britain and the Mediterranean, except that they called for a considerable amount of collaborative effort in moving stones that weighed many tons. They often came a great distance from where they were quarried. These stones were largely unfaced, and of irregular shape. It has to be admitted that we do not know the religious significance of the standing stones or menhirs. The dolmen – huge flat stones resting on a number of large, vertically positioned, flat supporting stones – were burial chambers that have lasted some 7,000 years, since the early Neolithic period. Why they had to be made of such large units we don't know: some durability can be achieved with smaller building blocks. To lift a thirty-ton block off the ground, and position it on top of uprights at a height of up to ten feet calls for equipment that is advanced – beyond the block and tackle available to us over the last few hundred years, and prior to twentieth-century earthmoving equipment.

Yet this is nothing compared with certain megaliths that are visible at several widely spaced parts of the world. Take Sacsayhuaman in pre-Inca Peru: this is one of the most beautifully built fortresses in the world. Many of its building blocks weigh as much as 300 tons; their surfaces are multifaceted, but fit together so neatly and tightly that a knife blade cannot be slid between them; and there is no mortar holding them together. How these blocks have been tailored to fit so precisely, no one knows. It has been suggested that they were positioned on top of each other, and then moved until the grinding action allowed each block to nestle tightly. As a solution, this is wholly unacceptable: firstly the blocks are far too heavy to agitate in this way – in any case this would not work when fitting together what amounts to the world's largest jigsaw puzzle; secondly, the pieces of rock that broke loose as a result of the grinding would remain in the interstices and prevent the close fit which is so readily apparent. No, these walls at Sacsayhuaman have been constructed by a method whose technology has long since disappeared. It seems almost as if the blocks have been roughly shaped, then plasticised by some unknown technique, and lastly fitted into position allowing the surfaces to flow together to form the tight bonding seen; any surplus plastic rock that had been squeezed from the joints would then have been cut away. The fact that joints are bevelled suggests that surplus material has been cut away at the meeting point of those blocks. This is a wholly speculative suggestion, of course, because we just do not know how these walls were assembled.[12]

The lower-level walls of the erstwhile Inca temple at Cuzco are made in the same way; later building – resulting in the present Christian cathedral – has topped these walls with Spanish architecture which does not possess any of the tight-fitting magic of the lower parts of the walls.

Apparently, when the Spanish conquerors were trying to adapt this temple they tried to destroy the lower levels of these megalithic walls, and were unable to shift them. It is possible that, once these blocks have settled together, they have effectively melded into one large entity; so separating them would amount to requarrying them.

So who built the walls? One thing is certain: the <u>Incas did not.</u> It's worth quoting the sixteenth-century

12  One of the guides in the Sacred Valley came up with a curious story. According to an unnamed German professor, there is a variety of Amazonian woodpecker that makes its nest in holes that it has dug in rocks. The bird achieves this penetration by chewing a specific leaf which enables it to plasticise the rock while pecking with its very powerful beak. The chemical in this saliva-assisted leaf apparently doesn't harm the woodpecker. This information needs a lot more substantiation before it can account for the building successes of the pre- Incas.

Spanish chronicler Garcilaso de la Vega on the fortress at Sacsayhuaman:

> It is made of such great stones and in such great number that one wonders simultaneously how the Indians were able to quarry them, how they transported them, and how they hewed them and set them one on top of the other with such precision. For they dispose of neither iron nor steel with which to penetrate the rock and cut and polish the stones; they had neither wagon nor oxen to transport them, and in fact there exist neither wagon nor oxen throughout the world that would suffice for this task, so enormous are these stones, and so rude the mountain paths over which they were conveyed.

He goes on to say that at a previous time an Inca king had attempted to copy the efforts of these earlier constructors. The king had set about bringing a huge boulder from several miles away to add to the structure at Sacsayhuaman and had employed 20,000 Indians to haul the rock up and down the neighbouring mountains. At a certain point the hauliers had lost control of the rock, which had fallen and crushed 3,000 of them. Obviously, the Incas the Spaniards encountered were only tenants of these amazing buildings, and not the builders.

So who were the skilled builders of these mysterious Altiplano edifices? Not only Sacsayhuaman and Cuzco, but Machu Picchu and Tiahuanaco also show examples of this astonishingly beautiful construction, which exhibits a technology that cannot be understood or explained by contemporary builders. Then, too, how many millennia do we have to go back to find a reason why these skills were lost?

**Tiahuanaco again**

We believe Tiahuanaco holds an answer to this question. There are unmistakable pointers to the city being in existence before the catastrophe that signalled the end of the last so called ice age. In other words, Tiahuanaco was functioning more than 11,500 years ago. The great expert on these astonishing ruins is Professor A. Posnansky of the University of La Paz, a man who has spent much of his life studying the place. The accepted view is that the earliest civilisation to inhabit the city existed around 300 BC. Since then four more groups of people have held sway. Posnansky believes that when Lake Titicaca was tilted, some 9,500 years BC, it and other lakes further up the Altiplano unleashed such a welter of water and debris that the earliest Tiahuanaco was buried to a depth of several metres. This original city is known, but has not yet been excavated; a team of Japanese archaeologists would like to begin digging, but Bolivian national jealousy stands in their way.

Posnansky believes the city was in existence 15,000 years BC. One of his main arguments is based on the alignment of the Kalasasaya, an observatory-like structure in the centre of the ruins. This building appears to have been used to fix the seasonal positions of the Sun. If this were so, the indications are that this building was put up at a time showing the obliquity of the ecliptic, some 17,000 years ago. We do not go along with this argument, as it does not fit with the understanding that the poles have shifted several times within that period. Nevertheless, it's quite conceivable that Tiahuanaco could be that age, supported by other evidence.

Carved on many of the stone blocks of these ruins are animals that are no longer found in South America. Elephants are shown; they have not been present for several thousand years, and conceivably not since the end of the Pleistocene, 11,500 years ago. Equally interesting is that Toxodon, a prehistoric three-toed mammal measuring some 2.75 metres in length, and 1.5 metres high at the shoulder, which disappeared at the end of the alleged ice age, is depicted on some of the building blocks, in independent three dimensional pieces of sculpture, and on pottery. This amphibious mammal, looking like something between a rhino and a hippo, was present in South America in vast numbers. Toxodon is only one of many types of animal depicted in Tiahuanaco which have been extinct since the end of the Pleistocene.

There are signs in these ruins that reveal the catastrophe which destroyed life and the city itself. At various places building blocks lie scattered like the work of an untidy giant. Lake Titicaca at that time came up to the edge of the city – where a megalithic harbour had been built. The remains of animals and humans have been found, according to Posnansky, in the same alluvia as marine fish, indicating that the altitude of the ruins was totally different from today.

The earthquake that tilted and spilt Lake Titicaca, lowering its shoreline by a hundred feet in places, has left the present edge of the lake some twelve miles from Tiahuanaco. It has also elevated the surrounding Altiplano to a height that has made cultivation nearly impossible. The disaster almost certainly occurred at the end of the Pleistocene.

We have every right to assume that this catastrophe prevailed worldwide. Our argument, stated many times, is that a pole shift probably caused the death of more than 80% of the world's fauna, including humanity. So is it surprising that the technology governing the construction of these amazing buildings in South America, and the knowledge required to move their vast stone blocks many miles from the point of quarry, has simply disappeared.

# PART II

## *There is Something Missing From the Solar System*

**Why won't the planets run out of energy?**

It is some three hundred years since a new theory explaining the way the solar sytem works came into being. The theory of gravity[13] by Sir Isaac Newton (1642–1727) has been accepted by the scientific world, and, though challenged by a new theory almost every year, has been largely found to work. It continues to be challenged because nobody can quite believe the astonishing fact that one force – gravity by itself – can govern the movements of all the bodies in the solar system.

In order to make his gravity theory work, Newton had to combine it with his first law of motion.[14] Thus it is that a planet, for example, travelling in a straight line will be made to constantly fall towards the Sun by the force of gravity pulling these two bodies towards each other. The planet will continue to fall towards the Sun as long as it has enough momentum to prevent it hitting the Sun. This momentum (mass x speed) is very

---

13 'Every particle of matter in the universe attracts every other particle with a force that varies directly as the products of their masses, and inversely as the square of the distance between them'
14 'A body remains at rest or maintains a uniform motion in a straight line, unless acted upon by a force.'

important; if a planet's momentum were to increase, that body would tend to travel more in a straight line and would fly off at a tangent to its present orbit. Likewise, if the momentum ran out, the planet would start to spiral in towards the Sun, and – it is believed – would hit that body. We know that in the case of manmade satellites launched into orbit round the Earth this momentum does eventually run out, and the satellite hits the Earth (or, in actual fact, is usually burnt-up on re-entry into the Earth's atmosphere). This loss of momentum should alert us to the likelihood that planets are suffering the same braking action – albeit infinitesimally slowly – assuming Newton's theory is correct.

As it happens, contemporary science chooses largely to ignore the possibility that planets could be losing momentum as a result of forces which are apparent. The view is that this loss of inertia is so small that it can be deemed not to exist: planets were given their inertia *ab initio*, that is, at the time when they broke away from the proto-Sun. They have not slowed down since. And this is why the theory of gravity does not bother about the momentum which is regarded as everlasting. So the only effective force governing movement in space is gravity, balanced by the somewhat artificial centrifugal force, which is not a real force but rather a derivative of momentum.

Why might this interpretation of solar happenings not be correct? Well, to start with, the solar system has been going for 4.6 billion years, or more. That is a long time for a system to keep going without any top-up of the original energy. Even one tenth of that time would be long. Under the accepted rules of how the system functions the planetary bodies should be slowing down. But we know this is not happening, so could there be something wrong with Newtonian theory? If that were the case, the momentum of the planetary bodies must be being maintained with energy from some other mysterious energy source.

Here are two very good reasons why a heavenly body should slow down. Firstly, the solar wind creates bow waves in a comet when the wind meets that body head-on in the latter's rush towards the Sun. This comet may or may not have a magnetic field round it, according to James Van Allen, but either way it creates a gaseous envelope when within the inner solar system. This envelope is compressed on the comet's leading edge by the solar wind, which then draws the tail out to those astonishing multi-million mile distances we have seen in Halley's and other comets. The retarding effect of this solar wind – together with light and other electromagnetic radiation – is readily apparent. The wind constitutes a continuous outpouring from the Sun, impacting all bodies in the solar system, and decreasing in strength as the inverse square of the distance from the Sun. Its effect on planets is equally apparent, though at a rather different angle, which will become an important topic later.

Secondly, irrespective of the acceleration and deceleration experienced by planets approaching or receding from the Sun in their elliptical circuits, the net effect of gravity at right angles to the tangent of a planet's orbit will be to remove kinetic energy from that planet. Professor Whipple (*Orbiting the Sun*) and others may disagree with this on the grounds that the planet, in resisting the pull of gravity with its centrifugal tendency, is returning to its innate state of rectilinear travel. Nevertheless, any body resisting a pull is required to give up energy to achieve that resistance. A child on a swing is a good example of this: without further input of energy, each 'high' must be lower than the previous one. Let's not forget, however, that at times the planet appears to be urged forward by the gravitational pull being at less than 90°, and subsequently retarded by the pull at more than 90°, depending on where it may be on its elliptical path.

We are, therefore, looking for this missing force that keeps planets going and enables them to resist the gravitational pull that would otherwise compel them to collide with the Sun. That this force should be some mysterious thrust *ab initio* several billions of years ago is not good enough. But wait – is that initial kinetic energy there anyway? If we look at Mercury, the planet which, apart from Pluto, has the most elliptical orbit, we see that it largely disregards that earlier given momentum. When closest to the Sun, its perihelion, the planet travels at 58 km per second; then in the astonishingly short period of 44 days at its aphelion it has slowed down to 38 km per second. This throws into doubt the whole idea of the thrust *ab initio* that all satellites of the Sun are supposed to have received. The point is particularly emphasised with comets on long elliptical orbits where the comet comes to a near standstill at its aphelion, prior to 'turning the corner' and making the long journey back towards the Sun. All right, 'gravity' assists the speedup when the satellite is approaching the Sun, and slows it down when retreating from the Sun, but there is no innate momentum in the comet's case, and not nearly as much as there is thought to be in the planet's case.

Clearly it's convenient to disregard this thrust *ab initio*. As we have suggested, conventional science does exactly that; and this leaves us with just one force doing all the work – gravity. But who ever heard of one force by itself controlling bodies in space, and even more strangely causing the planets to speed up and slow down. Moreover, that force, instead of pulling the planets in the expected direction, actually effectively acts on them almost at right angles to their line of progress. So how is it that a force in one direction causes heavenly bodies to fly in another direction?

There's something odd here. If only elliptical orbits were not the case, and planets travelled in perfect circles, it would be much harder to argue with Newton. That thrust *ab initio* would have established a constant speed, and might be the undisputed propellant of the planets.

Planets subject to this single force should, while trying to obey Newton's first law of motion that they travel in a straight line, behave differently. With gravity pulling them towards the Sun, and centrifugal force compelling them away from the Sun, they should travel in perfect circles. Oddly, they do not do this: all motion in the solar system is elliptical to a greater or lesser degree. To that extent the planets obey Kepler's first law, which empirically states this fact,[15] but does

---

15   Kepler's first law states that the orbit of a planet is an ellipse with the

not explain why this curious happening occurs. As previously stated, centrifugal force is that which balances gravity, but it must be repeated that centrifugal force is not a real force in its own right: if the momentum were not there, the centrifugal force would not be there either.

Let me recap with an analogy: if we take a stone on a length of string and swing it in a circle round our head, there are three forces acting on that stone to keep it on circuit. Firstly, the string acts like gravity and pulls the stone towards us; secondly, the same tautness of the string acts like the centrifugal force which is trying to take the stone off in a straight line tangential to the circle round which the stone is flying; and thirdly, our hand is propelling the stone ever onwards round our head. Without this last propellant being in place, the other two forces cannot continue to behave in the way described. We must find this missing force, even though we know – from the way comets and other bodies speed up and slow down – that this force is not a consistently stable one.

For some three centuries gravity has been given this preeminent role as, effectively, the only force governing movement in the solar system. It has been wearing the emperor's new clothes for far too long. It is time the emperor got dressed again.

**What is gravity?**

Compared with the other forces operating in nature – for instance, nuclear or molecular forces – gravity is very

---

Sun at one of the foci. When the two foci are at the same spot, the ellipse becomes a circle.

weak. It plays no part in the structure of matter, it only appears to influence the position that one piece of matter holds in attracting another. There are instances when that relationship no longer holds true. For example, when we fall from a height into water, we are initially compelled waterwards by gravity until we have passed through the surface of the water. Thereafter, gravity becomes the lesser force, and we float upwards to the surface. At this point density has become the controller of the moment: the density of our bodies is less than that of the surrounding water and we rise to the top. The role played by differing densities is often seen in liquids and gases: a hot-air balloon rises because the heated air, expanding its structure, takes up more space than the surrounding colder air. Its density is less, and it rises, disregarding gravity. Similarly, liquids of mixed density such as oil and water sort out their levels: the less dense oil goes to the surface, in spite of gravity. A lighter metal such as aluminium when dropped into mercury happily floats to the surface. Solids, too, would ignore gravity and adjust their densities if they could. Unfortunately, friction is such that movement between heavier solids on the surface of the Earth cannot readily take place spontaneously. If this friction were not present, we would see solid objects change places with each other, just as we see smoke changing place with denser, cold air. We are led to believe that within the mantle and core of the Earth, this adjustment has taken place, and the heavier solid iron has sunk to the centre of the core. So one of the components of gravity appears to be density.

On leaving the Earth's atmosphere the density component ceases to play a role. Gravity in its conventional form is thought to be the only force acting on bodies roaming the solar system.

It has been said, contrary to received thinking, that matter does not attract matter in inverse ratio to the square of the distance separating the two pieces of matter. In other words, once nuclear or molecular interaction has occurred between two pieces of matter, that matter is inert, and incapable of attracting other matter. This may or may not be the case; it may not seem of particular relevance. Certainly, within the atmosphere of a planet we have shown that part of the attraction is explained in a different way. All the same, we have to find out what is causing attraction at a distance between two bodies within the solar system. Magnetic fields attract or repel each other; they also attract matter in a way that makes it appear as though matter is attracting matter. Therefore it can be said that the vertical component of magnetism is the main constituent of what we call gravity.

All matter has a surrounding magnetic field, although in non-ferrous materials it's not so apparent: it is this part of 'gravity' that effects the relationship between bodies in space. We could be accused of being pedantic here, but the New Theory will show that this distinction is important.

Perhaps the most interesting thing about gravity – as defined by Newton – is that it does mathematically account for most of the movements in the solar system to within a very precise degree of accuracy. It explains what we see. So why are we bothering to look for the missing force which should accompany gravity, when to date no theory seems to work better, in practice, than gravity operating as a solo performer? The answer is that if we were to come up with a convincing new theory to explain movement in the solar system, it might not change the mathematical calculations of those movements, but it would give a startling new twist to the way we thought about the solar system. It would also cast a strange new light on the changeability of that system – and that is what the first part of this book is about.

**The New Theory**

The Milky Way is a vast magnetic field within which there are millions of secondary magnetic fields. At the centre of every solar system within the Milky Way lies such a magnetic field. In our case this Magnetic Centre (MC) is at the centre of mass of the solar system. The Sun is comparatively close to the Magnetic Centre, but the distance between them varies.

It is known that the Sun's movement with respect to the stars is irregular. But the conventional view is that this movement is wholly dictated by the gravitational perturbations of the planets. Sixty per cent of the movement is caused by Jupiter, and Saturn coupled with Uranus and Neptune takes the figure up to over ninety-five per cent. Without this planetary interference the belief is that the Sun would be static in relation to the solar system; in other words, no force would be acting on it, nor would it be moving spontaneously.

The New Theory says 'rubbish'. The Sun is in fact moving in an orbit of its own round the Magnetic Centre. Like all other bodies in the solar system it is on an elliptical course. It would be on a circular orbit if it were not perturbed by the planets. Any attempt to calculate the interference of nine planets is hellishly complicated and can only be achieved by a computer and with difficulty. The Sun is attracted by the MC, but is at the same time prevented from getting any closer to it by the pressure of its own solar emissions.[16] So, although received science considers the Sun to be the centre of the solar system, it is in fact not.

The movement of the planets comes about as a result of the interplay between the Sun as a repelling agent and the MC as an attracting agent. The Sun's radiation (mostly photons) to a greater extent, and the solar wind to a lesser, repel planets, while the MC attracts them. It is the torque action of these two push-pull forces that compels the planets and other bodies to rotate on their axes, and to travel round the solar system.

At this point many readers will ask themselves whether they should bin this book or read on. What evidence do we have for this outrageous theory? Can any theory be so extreme as to suggest that the Sun repels rather than attracts the planets? But wait – we know that there is pressure exerted by sunlight: barometric pressure varies <u>directly according</u> to the weight of sunlight reaching

---

16  These emissions can be divided into the solar wind, which is hot plasma or ionised gases travelling at the lowly speed of some 400–800 km/sec, and solar radiation, which is mainly photons and other electromagnetic radiation travelling at averaged speeds of 300,000 km/sec. The latter is much more important.

the Earth's surface. And we know the force of sunlight reaching the magnetosphere is far greater, but that the atmosphere dissipates much of that force.

Equally, we know that gravity is a very weak force. So if we persist in giving the Sun this capacity to both pull and push, we are putting it into a somewhat contradictory role. The solar wind (let us refer to all solar emissions, sunlight and plasma by this term) shows evidence of this pressure: as we have stated, a comet approaching the Sun becomes flattened on its leading edge; its tail gets pushed several million kilometres to the rear side of the comet.

Earth's magnetosphere is compressed to some 10 Earth radii 65,000 km) on the daylight side of the Earth; while on the night side the magnetotail is extended, by a mixture of direct push and boundary-layer drag to a distance in excess of 1,000 Earth radii (6,500,000 km). James Van Allen has stated that the solar wind has a negligible effect on the movement of large bodies such as planets. However, this is quite blatantly and observably untrue.

**Why can't we see the Magnetic Centre?**

The answer to this important question is that the MC is a highly condensed body of matter which may be only a few kilometres in diameter. It is a black hole. And a black hole is extremely hard to detect. Once thermonuclear reactions have ceased in a star, there is no known source of pressure that can support it. It has to start collapsing in on itself.

According to received science, the crush of gravity overwhelms all outward forces. According to Professor Michael Zeilik:

> No material can withstand this final crushing point of matter. The collapse cannot be halted; the volume of the star will continue to decrease until it reaches zero.
>
> The density of the star will increase until it becomes infinite.
>
> Before a mass reaches total collapse, bizarre events occur near it. As its density increases, the paths of light rays emitted from the star are bent more and more away from the straight lines going away from the star's surface. Eventually the density reaches such a high value that the light rays are wrapped around the star, and do not leave. The photons are trapped by the intense gravitational field in an orbit round the star.
>
> The escape velocity from the star is then greater than the speed of light. Any additional photons emitted after the star attains this critical density can never reach an outside observer. The star becomes a black hole.

It is believed that a star only a little larger than the Sun at the end of its working life will collapse to a body of some three kilometres in radius. It has the same mass as ever, so the density is staggering. A body of two solar masses collapses to some four kilometres in radius. And of course, the gravitational field is beyond belief in its intensity. This black hole is, as we said, the Magnetic Centre in the New Theory. It has a mass that is somewhat greater than that of the Sun; so the Sun orbits the MC. Professor Zeilik goes on to ask how the black holes are observed:

> With difficulty! Light emitted inside cannot get out . . .
>
> In addition, a black hole is small, only a few kilometres in size. You'll have a hard time seeing an isolated black hole. But a black hole near any mass might be observable. Matter falling towards a black hole gains energy and heats up . . . If heated to a few million kelvins or so, the material gives off X-rays.

X-rays are pointers to the presence of a black hole. But if a black hole has collapsed to the point where it has passed through what is called its Schwarzschild radius – the point at which concentration of mass has become so extreme, and the gravitational pull so huge, that X-rays would not escape from the body – then, naturally, the X-rays would not be observable. If, however, the X-rays have come from the Sun, they might be seen moving in the direction of the MC.

Lively discussions on the possibility of the Sun having a companion have taken place. An article by E.R. Harrison, an astronomer at the University of

Massachusetts, appeared in *Nature* (24th November 1977): in it he suggests that the wobbly motion of the Sun is characteristic of a star bound to a binary companion. If it exists it must be a dark object.

He goes on to say that obviously a companion star on closed orbit must lie close to the ecliptic plane, so as not to disturb the planets excessively. He concludes his article:

> Has the Sun a companion star? I find it hard to believe that a star so close can exist and yet remain undiscovered.
>
> On the other hand pulsar observations of extraordinary precision imply that it might exist, and therefore a search for a companion star is perhaps worth undertaking.

An astronomer from the University of British Columbia, Serge Pineault, supported Harrison in a follow-up article in *Nature* (26th October 1978). He felt that the pulsar data could not be due to the presence of a faint white, red or even black dwarf in closed orbit round the Sun. However, the presence of a neutron star or black hole was indeed possible. Such a dark object could have remained undetected until now, he writes, but could now be detected by satellite photography. A search would be worthwhile.

Interestingly, others have been thinking in this vein more recently. Jason Barker and James Larsen (blackhole@Yahoo.com) in January 2000 wrote:

One theory about black holes is that there is one supermassive black hole at the centre of every galaxy, and that is what holds all the different solar systems together. . . In a binary star system there are two stars present, and one of the stars may collapse into itself to form a black hole.

Received science generally believes that binary stars are separated by not more than 100 AU, or two and a half times the distance to Pluto. However, it is our belief that a black hole somewhat greater than the Sun in mass, and only a few kilometres in diameter, is located at the centre of the solar system, extremely close to the Sun. It would be responsible for intense 'gravitation', which the Sun would be fending off with the thrust from its solar wind and photon radiation. And that is just the relationship between the Sun and MC.

Should the reader find the existence of the MC – in the guise of a black hole – to be a postulate too far, let us approach this quest differently. Let's just assume, without acceptable supporting evidence, that the Magnetic Centre of the solar system exists. It is after all a vital fifty per cent component of the push-pull solar theory; and this theory provides answers to several unsolved mysteries. It explains orbital revolutions; it provides an understanding of planetary rotation (no other theory so far); and governing all else, it offers an explanation of the missing force that gravity needs to accompany it, the substitute for the mysterious thrust *ab initio*. If the New Theory holds up, the next step is to find the Magnetic Centre of the solar system.

**Sunspots**

Sunspots are well-known features of the Sun; occasionally they can be seen by the naked eye. They first received astronomical attention when Galileo turned his telescope on them in 1610. They are points on the Sun's surface that register much higher magnetic readings than their surrounds.

At the same time the local heat level is well below that of the ambient solar temperature. Sunspot numbers – and hence magnetic-field activity – increase and decrease over an eleven-year cycle. At its magnetic maximum the Sun causes storms in the Earth's magnetosphere, which at the Earth's surface lead to huge power surges and blackouts at power stations, and disruption of radio communications. Five and a half years later, at its minimum, the Sun's magnetic activity indicates a loss of heat at the Earth. This was made clear during a period known as the Maunder Minimum, between roughly 1640 and 1710, when there were very few, if any sunspots, and the Earth suffered a bout of extreme cold. Allegedly, the Earth's temperature worldwide fell by more than a degree centigrade.

So what do sunspots have to do with the New Theory? The answer is, a lot. The Sun is on an elliptical orbit round the Magnetic Centre. Every eleven years these two bodies reach perihelion: here the magnetic induction in the Sun reaches its maximum, as the MC's vast magnetic field comes closest. The sunspot count reaches its highest figure. Five and a half years later the sunspot figure reaches its lowest count, as the Sun and

MC reach aphelion. It must be remembered that the MC is located on the ecliptic, whereas the orbit of the Sun moves above and below the ecliptic. Like the Earth, the orbit of the Sun is not fixed, but precesses round the MC. Thus there are times when the proximity of Sun and MC at the time of perihelion is extremely remote. It is at this juncture, and for many perihelions before and after, that much less magnetic induction is initiated by the MC in the Sun; and this naturally causes a Maunder Minimum. Gradually, as the perihelion precesses towards the ecliptic, the distance between the Sun and MC reduces. There is then an increase again in the number of sunspots. Obviously, once more there is an increase in the solar energy reaching the Earth.

So where is the MC? Can we point to its location at any one moment? We know that the Sun takes eleven years to circle it. We know that Mercury is on an extremely elliptical orbit of 88 days' duration, indicating that this planet comes in much closer proximity to the MC than any other. This suggests that the MC lies on the ecliptic between the Sun and Mercury; if it were outside Mercury, the orbit of Mars would be more disturbed. But how close to the Sun is the MC? The problem is that it is very difficult to discover the Sun's speed in orbit. Obviously, the higher that speed the closer together lie these two influential bodies.

The search for the MC calls for observations of X-rays emitted by the Sun, before and after the solar maximum. At the same time Mercury's perihelion offers itself as another pointer to the location of this very small black hole.

**How the New Theory works**

As has been said, the MC attracts bodies within the solar system, but at the same time those bodies are repelled by photons and the solar wind. So planetary movement becomes an interaction between these two forces. The MC's attraction acts, in our case, on the centre of the Earth (illustration No. 1). The solar wind is more complicated. Firstly, it does not flow radially in a straight line – it has an anticlockwise (as viewed from the North Pole) bias, due to the rotation of the Sun in a similar direction. This means that the frontal edge of the particles leaving the Sun at any one moment – travelling in a spiral – impact the Earth's magnetosphere, not at right angles to its orbit, but at some 45° (illustration No. 2). The front pushes the Earth into an anticlockwise orbit, since it is prevented from retreating from the Sun by the pull of the MC. Secondly, the solar wind, which primarily impacts the Earth's magnetosphere at the magnetopause (illustration No. 3), will have more effect on the trailing edge of the magnetosphere, which lies to the east of the Earth–Sun line (as viewed from Earth). The net solar wind reaching that side is marginally greater than that reaching the west side. However, it is enough to add to the push sending the Earth yet further into an anticlockwise orbit.

The rotation of the Earth comes about somewhat differently: the MC plays the same role of holding the planet against expulsion, but the solar wind – largely radiation in the form of photons which have passed through the Earth's magnetosphere – makes direct contact with the surface of the Earth. The resulting torque causes the Earth to rotate, due to the greater presence of this radiation felt to the east of the Earth–Sun line. The much-reduced torque, where the leverage is little more than 6500 kms from the Earth's centre to the point of impact on the Earth's upper-atmospheric surface, explains the very low speed of the Earth's rotation – one revolution every 24 hours, or less than 1,700 kph at the equatorial surface of the Earth – while the much greater torque at the magnetopause accounts for the speed in orbit of over 107,000 kph. Here it is interesting to note that the major planets rotate much faster than the terrestrial planets, due to their greatly enlarged radii. At the same time, they orbit the Sun more slowly due to receiving less solar wind at their greater distances from the Sun. It is, of course, true of all the planets that the further out each one orbits the slower it travels, since the solar wind becomes steadily more dissipated, and less compelling.

There are exceptions to this planetary rotation. Venus and Uranus have suffered axial shifts which we have discussed in Part I. Venus has shifted its axis through 177°, so it is now counterrotating; however, the solar wind has slowed down this rotation to a pace whereby it takes 243 days to make one revolution. Soon it will stop revolving and start again in the normal prograde direction. Uranus has only recently toppled over and lies with its polar axis nearly on the plane of the ecliptic: it is still rotating fast – once every 17.9 hours – so it will take time to come to rest, prior to starting its prograde rotation with a new polar axis.

As has been stated, the Sun revolves on a very tight orbit round the MC. When the Sun lies between the

Magnetic Centre and the Earth, it obviously has more impact on that planet, which tends to move further away from the centre of the solar system; likewise, when the MC lies between the Sun and the Earth the power of attraction predominates, and the latter moves closer to the centre. It is this variation in the push-pull nature of the system that explains why none of the bodies in the Solar system can ever have a circular orbit – only ellipses are possible. If the Sun and MC were further apart, the eccentricity of the orbits would increase; and likewise if the planets were closer to the centre, the eccentricity would increase. This is what has happened in the case of Mercury, which has an eccentricity of 0.206. It is worth emphasising at this point that by contrast with the New Theory, the conventional wisdom should only allow the planets to have circular orbits (barring perturbations). Propelled by the thrust *ab initio*, drawn in towards the Sun by gravity, and restrained from collision by centrifugal force, the planets could only travel on circular orbits. But they don't, and we must take serious notice of this fact.

When the Sun moves round to the west side of the MC, you might feel that the Solar wind would start to impede the progress of a planet in its orbit. It is true that the solar wind is not so efficient at this location, but two features mitigate this problem: firstly, the Sun is very close to the MC, and secondly, the Sun always imparts more solar wind to the trailing edge of the planet, thereby impelling it onwards in orbit. It must be remembered that any anticlockwise movement of a planet is not simply the resultant of the push-pull of the MC versus the solar wind; it depends, too, on the rotation of the Sun, which leads to more solar wind reaching the rear of a planet – hence the added prograde push.

Another point of interest is the action the solar wind performs in controlling the Earth's upper atmosphere. Under conventional thinking the light, uppermost atmosphere could readily drift away from the Earth (the lower atmosphere is held down by the weight of air above). As we know, it does not stray away. It is held in place by the solar wind, which is constantly massaging the atmosphere back into place as the globe rotates.

**Planetary locations**

As a result of the push-pull interaction of the Sun and MC, we can see why those planets which are of larger volume and yet less dense lie at a greater distance from the Sun.

This larger volume, and in most cases more extended magnetosphere, means that the planet is more subject to the force of repulsion than that of attraction. So, too, we see why the denser and smaller terrestrial planets are nearer the Sun: their greater densities and smaller volumes mean that they are more dominated by the attraction of the Magnetic Centre, less responsive to the solar wind.[17]

The larger planets, as we have said, due to their greater diameters receive relatively more leverage from

---

17 Pluto is an anomaly which has only recently become a planet. It has not yet adjusted to the forces described above. Its orbit takes it to an angle of 17° to the ecliptic, with an eccentricity of 0.25. This is more eccentric than the orbit of Mercury, which is 'permanent', whereas that of Pluto will eventually adjust itself.

the solar wind at their surfaces; hence they tend to rotate faster than the terrestrial planets. Nevertheless, the fact that they are so much further from the Sun means that each succeeding planet receives less solar wind, and in consequence their speeds in orbit reduce as their distances increase. You might ask why Uranus and Neptune, which are said to be denser than Saturn, orbit further from the Sun. The answer will lie in an upward correction being found for the density of Saturn, and a reduced density for Neptune.

It's of interest to look at the specific orbital radius held by each planet in relation to its neighbour. Bode's Law has provided a useful approximation of planetary positions. J.E. Bode (1747–1826) produced not a physical law, but rather a handy method of calculating the distances of planets from the Sun. (He, in fact, did not spot the 'law', but rather popularised it. The 'invention' belongs to his fellow German, J.E. Titius.) Write down the sequence 0, 3, 6, 12, 24, etc.; to each number add 4; then divide the resultant figures by 10. Of the first seven answers (0.4, 0.7, 1.0, 1.6, 2.8, 5.2, 10.0), six closely approximate the planets' distances in astronomical units (AU = the distance from the Sun to Earth) and the 2.8 astronomical units fixes the distance of the asteroids.[18] Bode's Law holds for the seventh planet, Uranus, at 19.18 AU; however, it fails for the eighth planet, Neptune. Pluto appears where Neptune should be, namely at 39.67 AU, and Neptune is at 30 AU. Clearly there has been some interference

18

|  | Mer. | Ven. | Earth | Mars | Ast. | Jup. | Sat. | Uran. | Nept. | Pluto |
|---|---|---|---|---|---|---|---|---|---|---|
| Bode's Law in AU | 0.40 | 0.70 | 1.00 | 1.60 | 2.80 | 5.20 | 10.00 | 19.60 | - | 38.80 |
| Actual distance | 0.39 | 0.72 | 1.00 | 1.52 | - | 5.20 | 9.54 | 19.18 | 30.17 | 39.67 |

between these last two: at times Pluto's very elliptical orbit carries it inside that of Neptune.

We have explained how the planets are positioned at these distances from the Sun by the interplay between solar wind and magnetic pull. However, that's not all. Another important force impinges on each body: albedo – the reflectivity of a solar body's surface – plays a significant role. Sunlight bouncing off the planets' surfaces offers a force of repulsion from one planet to another. This, in an extremely complicated interaction of factors – mass, volume (hence density), albedo and distance – explains why each planet responds to the push-pull of the Sun/MC, and keeps a respectable, appointed distance between itself and the next planet.

**The plane of the ecliptic**

All planetary bodies should revolve around the Sun/MC in a prograde or anticlockwise direction (viewed from the North Pole) on a more or less precise plane – the plane of the Earth's orbit or ecliptic. The divergence from this plane is never more than 7° (which is the case with Mercury).[19]

There are a few, highly exceptional, satellite bodies that appear to defy the rules and revolve in a retrograde or clockwise direction. Four of Jupiter's satellites – Ananke, Carme, Pasiphae and Sinope – are examples. They are in a state of change, having just become satellites; they are on very eccentric orbits, and not holding well to the

---

19  This ignores Pluto, which is not yet established in a proper orbit.

plane of Jupiter's equator. Given time they will change; but we will discuss these anomalies later.

So why do planets and their satellites move in this ordered way? Again we turn to the relationship between the Sun and the Magnetic Centre. The plane of the ecliptic has the MC at its centre, and the orbit of the Sun round the MC is not strictly in the plane of the ecliptic: at times the Sun is north of the MC, during which time the solar wind will be forcing the planets southward in their orbits; likewise when the Sun travels south of the MC, the solar wind raises the planets in their orbits (illustration No. 4). You might object to this on the grounds that the distances are so great between the Sun and the planets, and the Sun is so close to the MC, that the cranking effect would not be adequate to orchestrate the planetary bodies in this way. You must bear in mind that these are the main forces affecting planetary motion, and as such a little goes a long way (forgive the metaphor). The constant pushing of the solar wind from these two different vantage points above and below the ecliptic means that the planetary bodies are always being shepherded towards the ecliptic.

Nearly as important as the movement of the Sun above and below the ecliptic is the way the solar wind flows from the solar corona. The *Ulysses* space probe has made some fascinating discoveries in this area since its launch in 1990. The European Space Agency devised this spacecraft to fly over the North and South Poles of the Sun: that was a difficult mission, because leaving the ecliptic was only achieved by using Jupiter as a staging post, whereby the craft was slingshot into the plane of the Sun's axis. Anyway, four years later the spacecraft passed the South Pole of the Sun at some 80°, followed by the solar North Pole in the second half of 1995. The point of this little story is that Ulysses discovered that the solar wind did not leave the Sun's corona at similar speeds at all latitudes. From the higher latitudes of each hemisphere the solar wind was largely developing in the coronal holes near the solar poles.

At these higher latitudes it would leave the Sun at speeds of up to 800 km/second. By contrast, the solar wind leaving the lower latitudes would be travelling at speeds as low as 400 km/second (illustration No. 5). This information is hugely supportive of the New Theory. The slower wind from the lower solar latitudes is that which travels predominantly in the plane of the ecliptic, shepherding the planets in orbit. Should these planets venture too far to the north or south of this plane, they encounter the higherspeed solar winds, which promptly push them back into their rightful ecliptic positions.

**The planets' satellites**

The satellites of the planets operate in a way that is analogous to that of their primaries. The two forces of attraction and repulsion are provided in the case of attraction by the magnetic pull of the equatorial bulges of the primaries, and in the case of repulsion by the reflected sunlight of those primaries.

Thus, just as the MC provides an elastic tether for a planet, so the planetary bulge provides such a tether for the satellite which is situated (with the exception of the Moon[20]) on the plane of the planet's equator. The albedo, or reflected sunlight, of a planet can be a considerable percentage of the light received from the Sun; it ranges from 90% in the case of Uranus, down to 16% in the case of Mars.

This sunlight is delivered to the satellite in varying quantities from all parts of the visible disc of the planet, but always from a greater distance than the point from which the equatorial bulge exerts its pull. Thus any tendency of a satellite to deviate from the equatorial plane is met by an increase of reflected sunlight, which compels it to return to its designated orbit on the equator (illustration No. 6).

This reflected sunlight differs from the solar wind in that it has no ionised plasma accompanying it. Nevertheless, it does a very efficient job of propelling the satellites in orbit, as a result of the planet's rotation. The planetary satellites orbit their primaries at different distances depending largely on their densities: the denser satellites are more governed by the magnetic pull of the planets' bulge, less influenced by the outward push of the reflected sunlight. Thus we see the Galilean Satellites

---

20  The Moon differs from the satellites of the large planets in that it is almost a planet itself. It does not hold to the Earth's equatorial bulge, but orbits at 5° to the ecliptic, rather than the 23.5° that would be required if it were orbiting the Earth's equator. It is governed partly by the push-pull coming from the centre of the solar system, and partly by reflected sunlight from both the Earth and the Moon itself.

of Jupiter with descending orders of density orbiting in ascending distances from the parent:

|  | Density (x$H_2O$) | Distance from Jupiter |
|---|---|---|
| Io | 3.53 | 422,000 km |
| Europa | 3.04 | 671,000 km |
| Ganymede | 1.93 | 1,070,000 km |
| Callisto | 1.79 | 1,880,000 km |

The very slight eccentricities of the satellites' orbits are a direct result of their primaries' orbits: when the planet has more sunlight from the Sun – a function as we have said of relative locations of the MC and the Sun – it increases its albedo. This imparts added repulsion or attraction to the satellite, distorting its orbit. The speed of revolution is governed by the speed of rotation of the planet. The vast majority of satellites do not revolve round their primaries faster than those planets rotate.

There are two notable exceptions to this rule: Jupiter's XIV revolves in five hours, while the parent rotates in nine and three-quarter hours; and Mars' Phobos revolves in seven hours forty minutes, while the planet rotates in some twenty four and a half hours. This anomalous behaviour indicates one of two things: either the satellite is sufficiently close to the parent to absorb an inordinate amount of reflected sunlight, or the satellite has recently[21] arrived from outside that planetary system at a high speed, and will be slowing down to a more warranted speed. We

---

21   Recent, in astronomical terms, may be several thousand years.

believe that the latter is more likely in the case of Mars' Phobos, since the albedo of the parent is poor.

Satellites revolving round planets in a retrograde or clockwise direction would fall into the last-mentioned category – they have recently arrived. They will be forced to slow down and revolve in the opposite direction. You will be tempted to say don't be silly; how can they come to a halt and then speed up in the opposite direction? You would be thinking in conventional terms. Within the New Theory there is nothing to stop a satellite coming to a halt; it will not crash into the planet, because the reflected sunlight would make quite sure that did not happen.

The exception to this is a low artificial satellite or one where the high density and close proximity are such that the magnetic pull far exceeds the push of reflected sunlight; in the latter case, the satellite – losing the slight help given to it by centrifugal force – will crash to its primary.

We previously mentioned the four outer satellites of Jupiter which revolve in this clockwise direction. They are emulated by Neptune's Triton and Saturn's Phoebe – the former at a much closer distance to the parent. All suffer from the same problem: being in conflict with the angular flow of reflected light from the primary. Rather than revolving with its support, their orbits have become erratic. Thus, Jupiter's four are on orbits with eccentricities up to 0.4, and deviate from the plane of the equatorial bulge by some 30°; Triton deviates from

Neptune's equatorial bulge by some 28°, and Phoebe is equally eccentric and deviated. The orbit of Triton is known to be altering, and, being the closest to its primary, it is experiencing change more rapidly than other retrograde satellites.

We must allude to a curious feature of the large satellites. The Moon, the Galilean Satellites of Jupiter, Saturn's Titan and in fact all the close-in, non-retrograde satellites of the planets are locked into a rotational period which is equal to one revolution of their primary. This has conventionally been described as a gravitational or tidal lock, which compels each satellite always to present the same face towards its parent. This is not the way that lock works: those satellites that are lying comparatively close to their parents are pulled towards the planet by the magnetic attraction of what is effectively the nearest point on the equatorial bulge, as we have stated. They are repelled, and propelled in orbit, by reflected sunlight which comes to them from all parts of the planetary disc that faces them: in this way they receive this sunlight equally on their leading edge and their rear edge. Thus the satellites are prevented from rotating. The further out the satellites orbit, the more inclined they would be to rotate in a non-locked way, since the distance separating the regions on the planet from which attraction and repulsion arises would become closer together.

We see this non-locked situation develop with the Sun and planets: the distances are far greater than with the planetary satellites, so that relative to distance the proximity of the Sun and Magnetic Centre creates a much more acute angle at the planet. There is, too, relative movement between these two points from which forces originate, as they revolve round each other. While with the planets and their satellites there is no such movement between the push-pull forces. The planet's source of reflected sunlight and its magnetic pull are stationary with respect to each other. Mercury's proximity to the Sun explains why it rotates in a near 'locked-on' mode.

**Comets**

Comets have extremely elliptical orbits – so extreme that there was an earlier belief that many were hyperbolic,[22] which would indicate that they originated outside the solar system. The present belief is that even the long-period comets, of which more than 600 have known orbits, originate in and return to the outer regions of the solar system. Approximately half of these long-distance travellers move in a retrograde orbit. They have not yet been tamed by the interactive forces of the Sun/MC. Their orbits may not lie anywhere near the plane of the ecliptic, since according to Jan Oort they originate in a spherical cloud (to which he gives his name) around the Sun, which has a radius somewhere between 20,000 AU and 100,000 AU.

The shorter-period comets behave somewhat differently: they may have started life as much-longer-period comets, but after many passes of the solar centre

---

22  That they had come in from interstellar space, would only be making one journey past the Sun, and would be returning to somewhere that was truly outside the solar system. They were not on orbits that were closed in the sense that we know now.

they have been pressurised into conforming more to the attraction/repulsion of the Sun/MC, in the way we have outlined. Thus they have moved nearer to the plane of the ecliptic, and the vast majority are passing the centre in a prograde direction.

Halley's Comet, however, has not yet got round to behaving in this corrected way. Its orbit will eventually become prograde. At the moment it only slows down by about a day in an orbital period of 76 years, so it will take some time to achieve this adjustment – always assuming the solar wind has not eroded it away before then.

The eccentricity of all comets is fascinating. Some of the longer-period orbits are so eccentric that they are nearly parabolic (illustration No. 7). In other words, a comet may start its sunward return travelling back on almost the same route that it took on the last part of its outward journey (of course, nearer the Sun it would depart from its outward route in order to circumnavigate the Sun in the same direction as its previous circuit). This means that before 'turning the corner' at the aphelion end of the outward journey, and starting the return trip, a comet would be all but stationary. That point would be where the effective thrust of the solar wind would have ceased, but the pull of the Magnetic Centre was still exerting influence. Analogously, a child's swing behaves similarly: the upward momentum of the child is eventually exceeded by the pull of gravity; the skyward movement stops, and the child returns earthwards to start on another cycle.

Received science believes that when a comet is at this point of aphelion it has just enough motion to direct its return journey on a route which guarantees that it will not hit the Sun. This is the belief that says that gravity is the single force that compels bodies to go on falling indefinitely towards the Sun, and yet never hitting that Sun. In the case of the planets on near-circular orbits, and assuming a momentum *ab initio* (from somewhere near 4.6 aeons ago), it is just possible to believe that the system works in this conventional way. But with comets this belief becomes much harder to sustain: a comet coming towards the Sun from a near-standing start should, by Newtonian thinking, hit that Sun. It has little centrifugal force with which to avoid the Sun. Moreover, the distance from aphelion is so great that the Sun would have ample time to pull the comet towards it and correct any chance of the comet missing. Clearly, the Newtonian rules have a challenge to answer here.

In general, comets and other planetary bodies do not collide. This is not, in most cases, because centrifugal force compels them to miss the magnetically attracting body, but because solar wind or reflected sunlight repels them. However, it is possible for collisions to occur, as we saw in July 1994, when five fragments of comet Schumacher-Levy hit Jupiter in quick succession. Everything depends upon the speed of the approaching comet, and its direction. In the case of Schumacher-Levy, if it had been travelling directly towards Jupiter from a considerable distance, Jupiter's reflected sunlight rotating in a prograde direction would divert the comet before it collided with the planet. If however the comet had been travelling towards Jupiter on a path that would take it slightly to the east of the planet, and it had been moving at a great speed, the reflected sunlight may have been sufficient to deviate the comet, which would then hit Jupiter. Much hangs, too, on the density of approaching bodies: a dense object would be less responsive to the reflected sunlight from the planet. In summary, sunlight met head-on would not deviate the comet, but met at an angle would cause the comet to change course.

### Spacecraft *Pioneer 10* and *11*

In 2002, fascinating information was released on the behaviour of the two veteran spacecraft *Pioneer 10* and *11*. These two probes were launched by NASA in 1972 and 1973. Their targets were Jupiter, followed by Saturn, and then after passing the three outer planets to head for the stars. By this year, *Pioneer 10* had reached a solar distance of some 80 AU. NASA's Jet Propulsion Laboratory was still keeping an eye on them in case they revealed the presence of a planet beyond Pluto, but in essence the mission had ceased after passing Saturn. Astonishment followed the discovery that the two spacecraft were not where they were expected to be. Although twenty-four billion kilometres apart, each had slowed down: *Pioneer 10* was some 400,000 km short of where it should have been. What was causing this deceleration?

It seems that the observing scientists were aware of the pressure of sunlight on the probes, and had included it as a factor in their calculations. However, they were not aware that the pull of the Magnetic Centre would become such a dominating force when the solar wind had effectively 'run out'. To all intents, these spacecraft have become comets.

As the reach of the MC is greater than the solar wind, these craft will eventually all but stop, turn round, and return to the centre of the solar system. What scientists have failed to realise is that initially the balance between the magnetic pull of the centre and the solar-wind push (actually mainly pressure of sunlight at greater distances) would have been near parity, leaving the initial thrust of the probes to be the propellant. Now, with the magnetic pull becoming the main force, the spacecraft will step up their deceleration.

NASA scientists are casting around for the presence of some form of 'dark matter' to be compounding the gravity that retards these vehicles. What they fail to realise is that a black hole (the Magnetic Centre) is not only the 'dark matter', but 'gravity' as well, with the Sun playing a diminishing role in the reverse direction. So while NASA scientists thought their probes were heading for the stars,[23] they have in fact turned into comets. *Pioneer 10* and *11* will surely be heading back towards the Sun, but it may take them several hundred years to get there.

## Asteroids

The asteroids – a mass of minor planetary bodies – are in the main located between Mars and Jupiter. Some 2,000 have been discovered, some of which travel

---

23 They assumed that Newtonian gravity attracts with a force that is inverse to the square of the distances.

in pairs. They vary in size from the thirty largest ones, which have diameters of over 200 km, to others which are boulder-size. (Ceres, the largest-known minor planet, has a diameter of 940 km.)

These bodies travel in groups on orbits that are in the main parallel to one another. For example, a group on identical or parallel orbits crosses the orbit of the Earth: this results in asteroids/meteoroids hitting the Earth's atmosphere (meteors) or even being large enough to reach the Earth's surface (meteorites).

Now, the single most fascinating feature of the asteroids for us is that there is no relative force of repulsion acting on these bodies – in other words, keeping them apart – yet they do not collide under the action of gravity. Of course, in conventional thinking they are subject to the centrifugal force preventing them from falling into the Sun, but there is no centrifugal force separating one asteroid from another, so why does gravity not compel them to coalesce into one big planet between Mars and Jupiter? *Because gravity, as expressed by received science, does not exist.* Any linking up that may occasionally happen is only due to magnetic attraction, where that attraction is stronger than the push of reflected sunlight from the two bodies.

**Summary**

1. We have seen how the two push-pull forces of the solar wind – both ionised plasma, and solar radiation consisting of photons and other radiation – on the one hand, and the Magnetic Centre on the other,

govern all planetary revolutions and rotations in the solar system.
2. We have seen, too, how the magnetic pull of the planets' equatorial bulges interacts with reflected sunlight to govern the movement of the planetary satellites.
3. We have explained how the Sun, by moving in an orbit which is slightly inclined to the plane of the ecliptic, shepherds its flock to move in orbits close to the ecliptic. We have also shown how the faster solar wind from the Sun's higher latitudes helps to contain the planetary bodies which may try to stray from the ecliptic.
4. We note the parallel to this in the way the reflected sunlight of a planet holds its satellite on a single equatorial plane.
5. We have shown how the interaction of solar wind and magnetic pull impinge on the densities of planets to determine how far out from the primary those followers will locate their orbits.
6. We have pointed out the distinction between the speeds of revolution and rotation due to the different degree of leverage exerted at different points on a planet's system.
7. We have forecast the changes that will eventually overtake the retrograde satellites – they will stop orbiting, change direction, and begin to revolve in a prograde direction (but not in our lifetime!)
8. We have talked of the unruly comets of the outer solar system, and indicated how they must ultimately come to order, adopting less eccentric prograde orbits.

Many of these points are ignored by received science, or deemed to be explained as part of a system inherited from the start of the solar system some 4.6 billion years ago. This assumes little change in the relationship between heavenly bodies. Gravity, with help from the not-quite-real centrifugal force, does all the work. It speeds up bodies on elliptical courses, then slows them down. It keeps bodies on almost circular orbits going, yet it is tugging at right angles to the tangent of those orbits. All considered it is accorded magical powers: and amongst them is the ability to ignore entropy, and permit planets to behave like perpetual motion machines.

**Conclusion**

The New Theory sees the solar system through new eyes. It is no longer so concerned with how the system came into being; the theories of Kant or Laplace, although interesting, do not provide an explanation of where the planets are, or why they are moving in their present modes. We have all the explanations we need from the attraction– repulsion activity of the Sun–Magnetic Centre. Thus we have a much more flexible solar system than before: planets can far more readily change their positions if they are not dependent on the location and thrust given to them at the time the system was created (see Appendix).

As examples of this we should look at the strange angles of inclination to the ecliptic exhibited by the planetary axes: from 97.9° in the case of Uranus to 177° (retrograde) for Venus. Clearly there has been much change since the start of the solar system. So it seems that planets can readily tip over and reverse their direction of rotation. Equally, retrograde comets starting from the Oort cloud at the far reaches of the solar system become tamed. They lose some of their eccentricity; they pass the Sun on a prograde orbit; and their orbital planes get closer to the ecliptic.

So the need for fast movement by a body within the solar system is largely removed. Newtonian thinking demands that the gravitational pull of the Sun compels bodies to keep moving in order to generate sufficient centrifugal force to avoid colliding with the Sun. The New Theory changes all that: because the push-pull of the Sun Magnetic Centre creates a balance, solar bodies do not have to be perpetually moving in the *ab initio* way. Instead, the balancing forces are in place, making collisions something of a rarity.

In theory a body in the solar system could remain all but stationary in terms of the Sun/MC – without crashing into another body – until the interplay between these two had slowly developed that body's orbit. This stationary state would be unlikely to last long because perturbations from other planets would soon make themselves felt.[24]

In summary, we are now looking at a world that is not dependent on the angular momentum – either in the revolution or rotation of bodies – attributed to it by conventional theory. The kinetic energy of the planets

---

24  Perturbation no longer means gravitational pull. It now means magnetic pull or the push generated by reflected sunlight.

would have been expended aeons ago if it were not for the perpetual add-on energy received from the pull-push imparted to them from the centre.

This means that if, by chance, one or other of these two central forces were to change, or if the way in which they were received by the planets were to change, dramatic alterations would occur in the planets' movements. It is this new thinking that will lead us to realise that a planet's rotation does not have anything like the gyroscopic stabilising characteristics of angular momentum that we are taught. Instead a planet can be likened to a child's hoop: provided the child applies its stick to the perimeter of the hoop in a steady even way, the hoop will continue to roll; uneven or reduced application of this force will cause the hoop to wobble and fall.

By now the reader should have seen the importance of linking Parts I and II into the same book. The doomsday illustrated in Part I could not come about if the uniformitarian thinking of the nineteenth century pertained. Yet evidence of catastrophe is rife in so many aspects of the Earth's history, as we show. It has taken the second part of the book to reveal just how changing and vibrant the solar system is in constantly feeding forces to its followers – a system so far removed from the one force world of Newton.

Sadly, we have to accept that certain features of the New Theory lead to Pole Shift – and that means horror.

**Evaluating the New Theory**

The big problematic feature of the New Theory is the very existence of the Magnetic Centre. There is scant evidence for it from observation. However, from a functional viewpoint the Sun/MC relationship makes far more sense than the Newtonian belief that gravity does all the work. Newton assumed a solar 'big bang' that set the planets revolving indefinitely. It is known that their orbital speeds are not reducing.

These maintained speeds led scientists to calculate that there is much less resistance in the solar system than there actually is. Newton's theory makes no attempt to 'top up' the initial momentum that was originally imparted to those planets 4.6 billion years ago. His missing thrust is as serious an omission as the 'missing' Magnetic Centre.

But let's look at the blessings that the New Theory does bring with it. It explains why all bodies in the solar system continue to orbit the Sun. It explains why planets rotate – no other acceptable explanation has been found. It explains why the bulk of solar bodies hold to or near the plane of the ecliptic – that this is due to the start of the solar system is not good enough, since some of these smaller bodies are revolving in a retrograde direction, and far off the plane of the ecliptic. It explains why the giant planets and the terrestrial planets are revolving where they are – no other theory has put forward a reason for this. There are several more occurrences that find explanations in the New Theory, where none was previously available – speeds of revolution, speeds of rotation, and lock-on being but three.

### The solar system gives out a new feeling

We are used to thinking of the solar system as a rigidly fast-moving collection of planets, their followers and other more independent bodies. All depend for their independent existence on their fast motion, for without that they would collide with their primaries. Now all that has to change: speed is not important – it's no longer necessary to generate centrifugal forces to combat gravity. The solar system becomes a significantly more gentle, easy-going, and laidback place: it has taken the whizz out of the planetary bodies. So, without this innate momentum and with a constant need for input from the Sun/MC, planets become noticeably more fragile entities. The way they receive and react to the solar wind becomes much more significant.

Likewise, the solar wind has many differing ways it can deliver its energy to these planets. We are in a more changeable world than we ever suspected. It is those changes – some very dramatic, and with devastating consequences – that we have examined in the first part of the book.

In the first part we concerned ourselves with the planet Earth. We have now taken the New Theory and applied it to the way the Earth has to behave. The resulting deductions have made strange reading: the steady state deemed so unshakeable by twentieth-century scientists has been challenged. Catastrophe theory, so beloved by early nineteenth-century scientists prior to Lyell and Darwin, could have a wonderful new innings.

# *Appendix*

## The origin of the solar system

There is, to date, no satisfactory theory of how the solar system started life. The main problem is that no theory can explain why almost all angular momentum of the system lies in the planetary bodies, and less than one per cent is left with the Sun. The New Theory takes care of this problem: as we have pointed out, there is much less angular momentum in the planets than has been thought – their innate momentum counts for little, and they are dependent on the Sun/MC for a continuous energy input.

Modern science agrees that the majority of the bodies making up the solar system are of the same origin as the Sun. These bodies started life some 4.6 billion years ago, although the Sun may be slightly older. Why a solar system originates is a cosmological question beyond the bounds of this book. That there are magnetic fields at the centre of other solar systems is known; until now such a magnetic field has been attributed to the Sun itself, in the case of our solar system.

The fact that most of the planetary bodies behave in an ordered way – revolving round the Sun in the same direction and adhering approximately to the plane of the ecliptic – has led science to assume that the condensing nebula, out of which the system formed, was a single rotating body. It was from such a contracting interstellar cloud that Laplace (and in a similar way Kant) formed their hypotheses which showed a 'breaking off' of rings as the nebula collapsed in on itself. These rings became the subsequent planets. We refer to these theories because they are generally the most accepted of the many devised. All theories have been found wanting in that none have come up with a satisfactory theory that explains the misplaced angular momentum. Now that we have found an answer to this problem, the Laplace theory looks the nearest satisfactory one to follow.

Once the solar rings have broken free of the newly formed and contracting sun, they become subject to a solar wind and start to form themselves into spherical bodies which then rotate in a prograde way. This argument only applies, of course, once the Sun has coalesced to a size and state where it becomes incandescent, revolves round the Magnetic Centre, and emits the solar wind.

# *Bibliography*

Agassiz, Alexander. *Proceedings of the American Academy of Arts & Sciences*, 1876.

Allaby, Michael & Lovelock, James. *The Great Extinction: What killed the Dinosaurs and Devastated the Earth*, Secker & Warburg, 1983.

Allen, James Van. 'Magnetospheres, Cosmic Rays and the Interplanetary Medium', in *The New Solar System*, edited by J. Kelly Beatty and Andrew Chakin, Sky Publishing Corp, Cambridge, Mass., 1990.

Anastasi IV papyrus.

Barrow, R.H. *Plutarch and His Times*, 1967.

Bellamy, H.S. *Built before the Flood: The Problem of the Tiahuanaco Ruins*, 1947.

Bode, J.E. 'Titius – Bode Law'. *Encyclopaedia Britannica*, 1988 edition.

Brunhes, Bernard. 'Recherches sur la direction d'aimantation des roches volcanique', *Journal de Physique Theorique et Appliquée*, Series 4, 5, 1906.

Buckland, William. *Reliquiae Diluvianae*, 1823.

Chamberlain, R.T. 'The Origin and History of the Earth' in *The World and Man*, edit F.R. Moulton, 1937.

Darwin, Charles. 'On the Origin of Species', *Encyclopaedia Britannica*, 1988 edition.

De la Vega, Garcilaso. *La Florida del Ynka*.

De la Vega, Garcilaso. *History of Peru, Part I: 1608. Part II: 1617*.

Donn, William, L. *A Theory of Ice Ages*, Science, 1956.

Donnelly, I. *Atlantis, The Antediluvian World*, Sidgwick & Jackson, London, 1970.

Eddington, Arthur. *The Nature of the Physical World*, Cambridge University Press, 1953.

Einstein, Albert. Correspondence with Charles H. Hapgood: Albert Einstein Archives, The Jewish National University Library.

Elsasser, Walter, M. 'Dimensional values in magnetohydrodynamics' in *Physical Review*, 1954.

Emiliani, Cesare. 'Pleistocene Temperatures' in *Journal of Geology* 63, 1955.

Epstein, Samuel, Buchsbaum, Lowenstam and Urey. 'Carbonate-Water isotopic temperature scale', Geographical Society of America, 1951.

Ericson, D, Ewing, M., Wollen, G., and Heezen, B.C. 'Atlantic Deep-Sea Sediment Cores', Geological Society of America, 1961.

Ewing, Maurice: 'A Theory of Ice Ages', *Science*, 1956.

Folgheraiter, Giuseppe. *Rendi Conti dei Licci*. 1896. Archives des Sciences Physiques et naturelles (Geneva) 1899.

Galilei, Galileo. 'Letters on Sunspots' translated by Stillman Drake in *Discoveries and Opinions of Galileo*, Anchor, New York, 1957.

Gaskell, T.F. & Morris, Martin. *World Climate; Weather, The Environment and Man*, Thames and Hudson, 1979.

Gold, Dr Thomas: 'Instability of the Earth's Axis of Rotation', *Nature 175*, 1955.

Gudmansson, A. 'Geomagnetic Excursion in late glacial basalt outcrops in South Western Iceland', *Journal of Geophysical Research*, 1980.

Hapgood, Professor Charles. *Earth's Shifting Crust*, Chilton, Philadelphia, 1958.

Hapgood, Professor Charles. *The Path of the Pole*, Chilton, Philadelphia, 1970.

Harris, Papyrus.

Harrison, E.R. *Nature*, 24/November 1977.

Harwood, J.M. and Malin S.C.R. *Nature*, 12/February 1976.

Heim, Arnold, and Gausser, August. *The Theme of the Gods*, 1939.

Heirtzler, J.R. *The Paleomagnetic Field. The Floor of the Mid-Atlantic Rift*. 1975.

Herivel, John. *The background to Newton's Principia*, 1966.

*Herodotus. Book II: History of Egypt. Encyclopaedia Britannica.* 1988.

Hibben, F.C. 'Evidence of Early Man in Alaska', *American Antiquity VIII*, 1942.

Howarth, Sir Henry, *The Glacial Nightmare and the Flood*, 1893.

Hoyle, Sir Fred. *Ice. How the next ice-age will come – and how we can prevent it*, Hutchinson & Co. 1981.

Hutton, James. 'Theory of the Earth', 1795. *Encyclopaedia Britannica*, 1988.

Imbrie, John and Katherine. *Ice Ages; Solving the Mystery*, Macmillan Press Ltd., 1979.

Ipuwer Papyrus.

Jacobs, J.A. *Reversals of the Earth's Magnetic Field*, Cambridge University Press, 1994.

Jeffreys, H. *The Earth, Its Origin, History and Physical Constitution*, 1929.

Jowett, Professor B. *Translation of Dialogues of Plato, Vol. III & IV*, Clarendon Press, Oxford, 1875.

Kepler, Johannes. Kepler's Laws of Planetary Motion, 1609 and 1618. *Encyclopaedia Britannica*, 1988.

Kukla, George. 'Loess stratigraphy of Central Europe' in *After the Australopithecines*, 1975.

Lamb, Professor H.H. 'Climate: Present, Past and Future' and 'The Changing Climate' in *Selected Papers*, Methuen London 1966.

Laplace, Pierre-Simon. 'The system of the World', *Encyclopaedia Britannica*, 1988.

Lyell, Sir Charles. *Antiquity of Man*. 1863. *Life, Letters & Journals*, 1881.

Malin, S.C.R. *Nature* 12/February 1976.

Markham, Sir Clements. *The Incas of Peru*, 1910.

Matuyama, Motonori. *On the direction of the magnetisation of basalt in Japan, Tyosen and Manchuria*, Imperial Academy of Japan Proc.

Mercanton, P.L. *La Methode de Folgeraiter et son role en geophysique*, Archives des Sciences Physiques et Naturelles, 1907.

Merriam, J.C. *The Fauna of Rancho La Brea*, University of California, 1911.

Milankovic, Milutin. *Memories, Experiences and Perceptions from the years 1909–1944*, Serbian Academy of Science.

Newton, Sir Isaac. *Principia Mathematica*, 1687.

Pedrosa, Osvaldo. Originator of the New Theory.

Pineault, Serge. *Nature* 26/October 1978.

Plato. *Timaeus*, translated by B. Jowett, Clarendon Press, Oxford, 1865.

Posnansky, Professor A. *Tiahuanaco, the Cradle of the American Man*, 1945.

Pratt, J.H. 'On the Attraction of the Himalaya Mountains upon the Plumline in India', Transactions of the Royal Society, Vol. CXLV, 1855.

Prestwich, Professor Joseph. 'The Raised Beaches and Head or Rubble-Drift of the South of England', *Quarterly Journal of the Geological Society*, 1892.

Runcorn, Professor S.K. 'The Earth's Magnetism', *Scientific American*, September 1955.

Schott, Walter. *Foraminifera in the equatorial part of the Atlantic Ocean*, 1935.

Seneca, Lucius Annaeus the Younger. *Naturelles questiones*, translated by A. Geike and J. Clarke, 1910.

Serres, Marcel de. Note sur de nouvelles breches ossenses decouvertes sur la montagne de Pedemar, 1858.

Sernander, Rutger. *Klimaverschlechterung, Postglaciale*, ed. Max Ebert, 1926.

Tarling, D.H. *Palaeomagnetism*.

Tarling, D.H. *Continental Drift*, G. Bell, London, 1971.

Thomson, Sir William, alias Lord Kelvin. *Treatise on Natural Philosophy*, 1987.

Wadia, D.N. *Geology of India*, 2nd edition, 1939.

Wallace, A.R. Letter to Darwin, 1858, *Encyclopaedia Britannica*, 1988.

Warlow, Peter. *The Reversing Earth*, J.M. Dent & Sons, 1982.

Wegener, Alfred. *The Origin of Continents and Oceans*, 1924.

Whipple, Fred, L. *Orbiting the Sun, Planets and Satellites of the Solar System*, Harvard University Press, Mass.

Wright, G.F. *The Ice Age and its bearing upon the Antiquity of Man*.

Zeilik, Professor Michael. *Astronomy, The Evolving Universe*, 8th Edition, John Wiley & Sons Inc., New York.

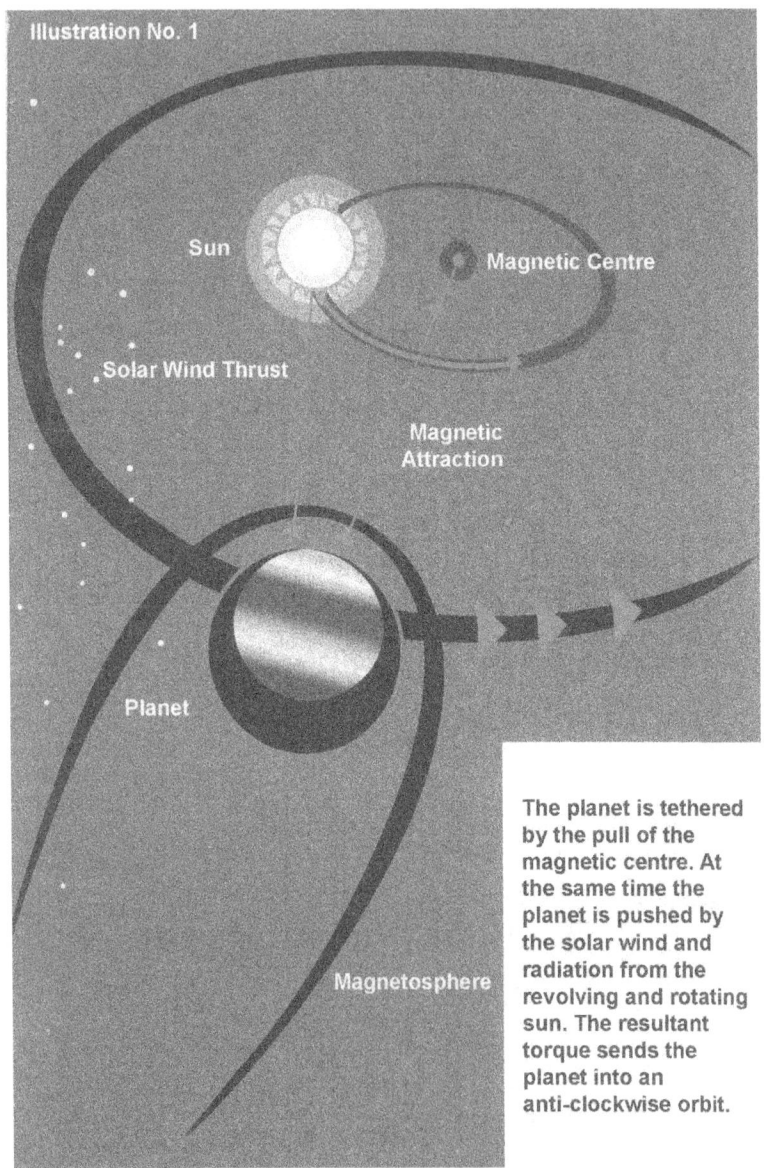

Illustration No. 1

The planet is tethered by the pull of the magnetic centre. At the same time the planet is pushed by the solar wind and radiation from the revolving and rotating sun. The resultant torque sends the planet into an anti-clockwise orbit.

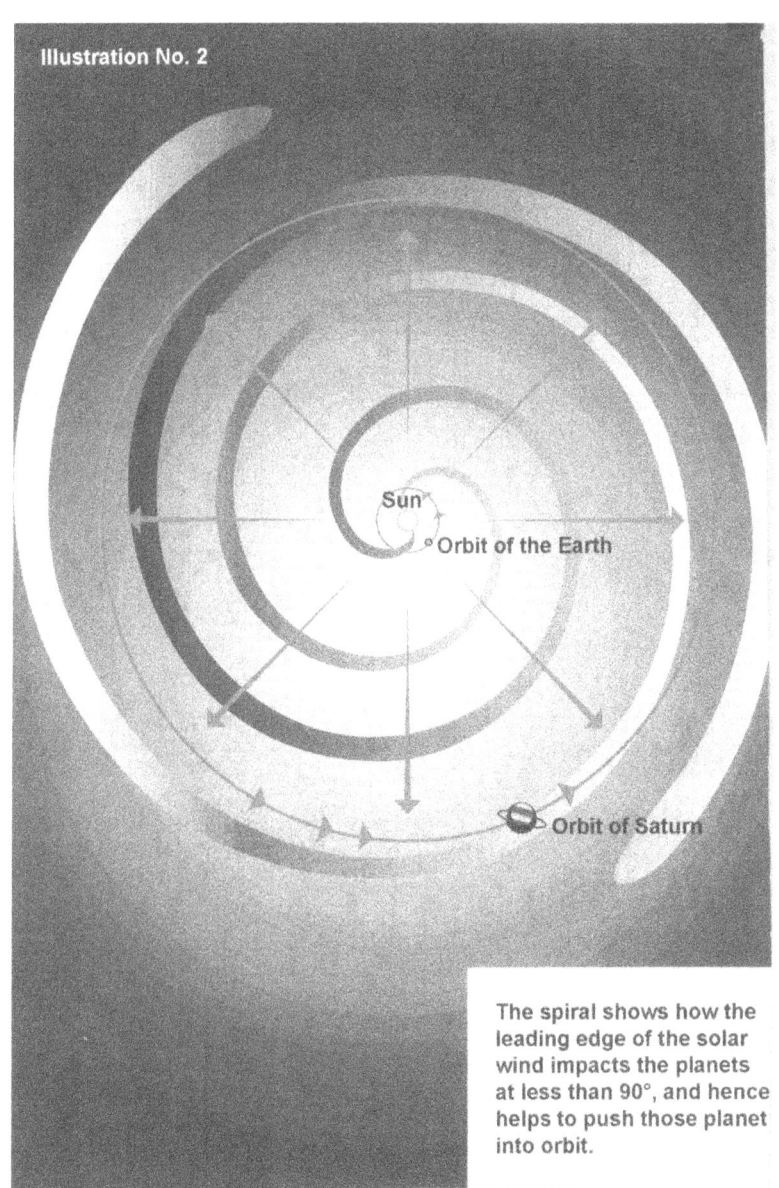

The spiral shows how the leading edge of the solar wind impacts the planets at less than 90°, and hence helps to push those planet into orbit.

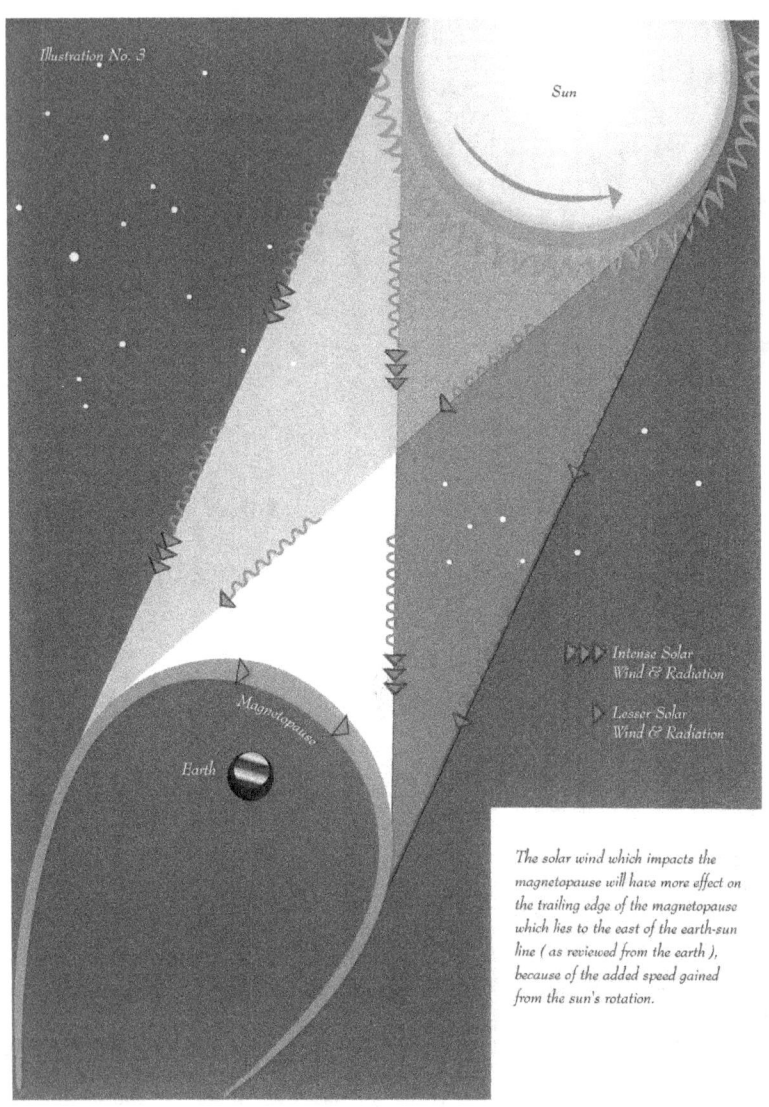

The solar wind which impacts the magnetopause will have more effect on the trailing edge of the magnetopause which lies to the east of the earth-sun line (as reviewed from the earth), because of the added speed gained from the sun's rotation.

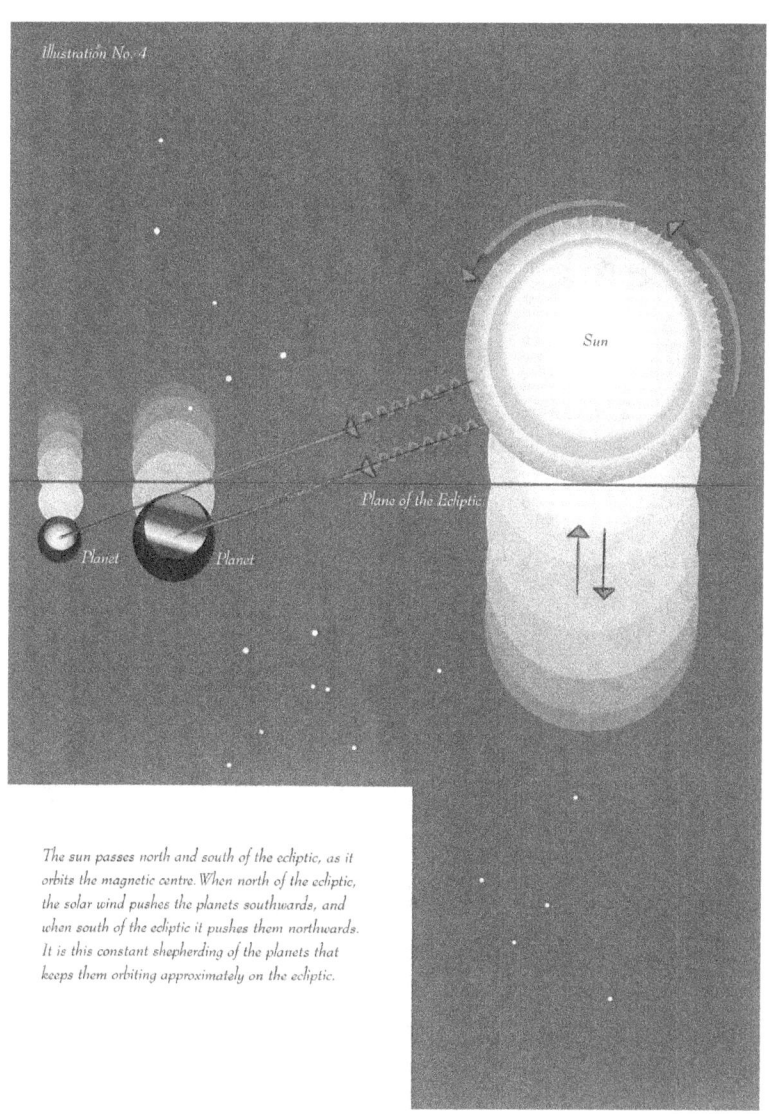

*The sun passes north and south of the ecliptic, as it orbits the magnetic centre. When north of the ecliptic, the solar wind pushes the planets southwards, and when south of the ecliptic it pushes them northwards. It is this constant shepherding of the planets that keeps them orbiting approximately on the ecliptic.*

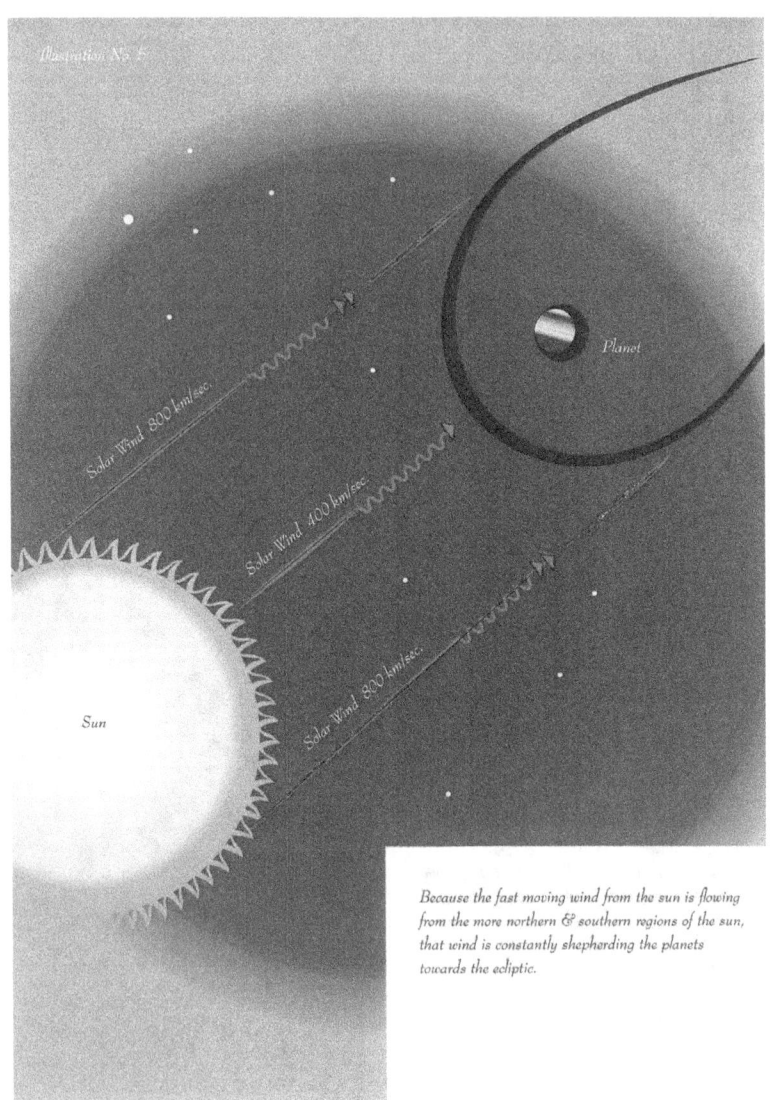

Because the fast moving wind from the sun is flowing from the more northern & southern regions of the sun, that wind is constantly shepherding the planets towards the ecliptic.

Illustration No. 6

Satellites of the planets are tethered by the magnetic pull of the planets' equatorial bulge. At the same time, they are repelled by sunlight reflected from the planets' presented disc.

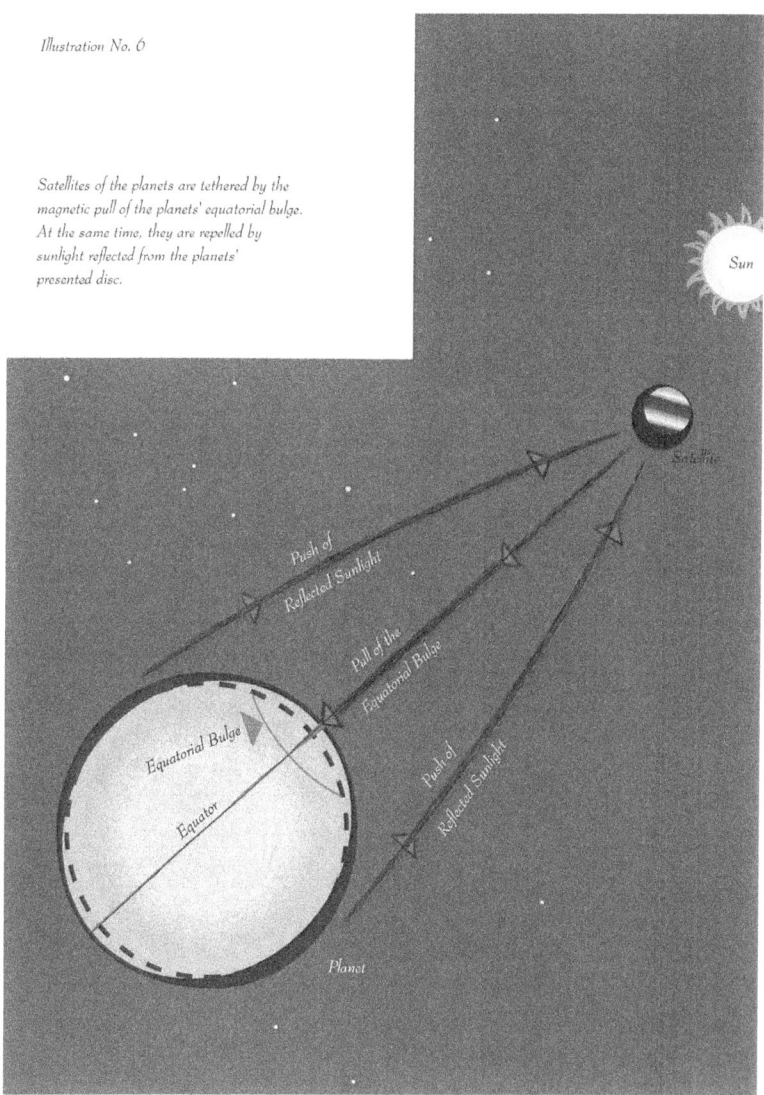

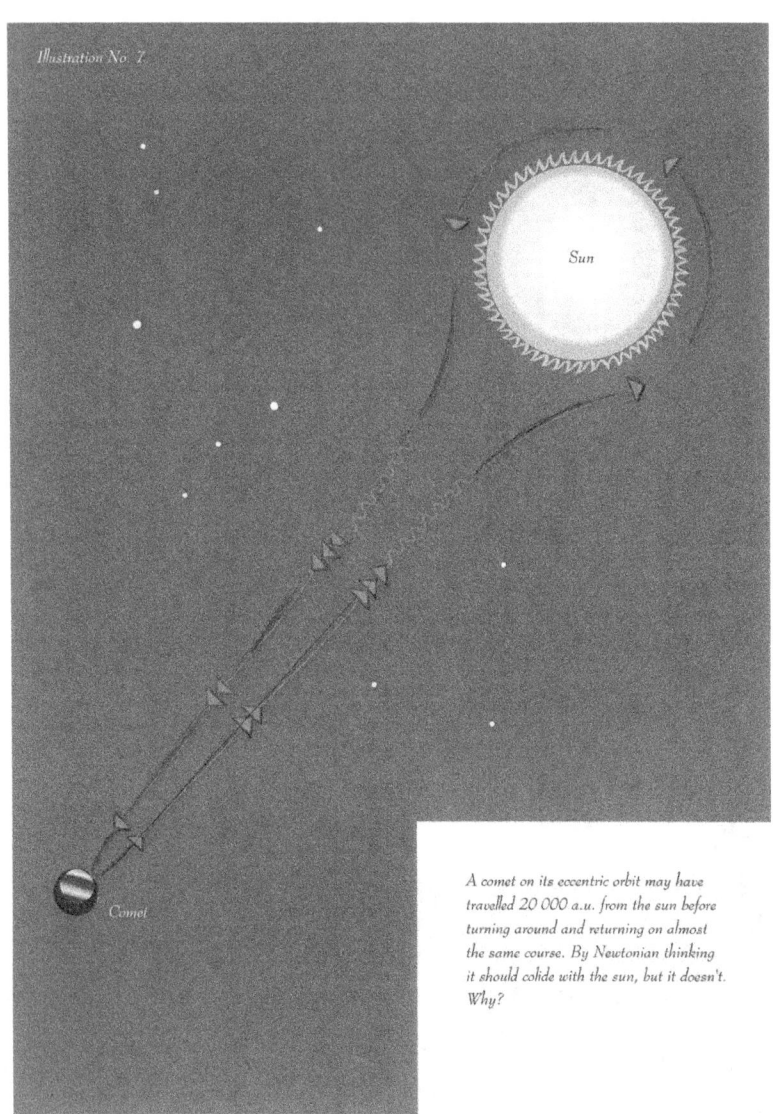

A comet on its eccentric orbit may have travelled 20 000 a.u. from the sun before turning around and returning on almost the same course. By Newtonian thinking it should colide with the sun, but it doesn't. Why?

*The Great Day of His Wrath* by John Martin

# *Index*

## A

Agassiz, Alexander  14
Agassiz, Louis  46
Alaska  8
Andes  13
animal remains  15, 69
   bones  3, 8, 9, 10, 11, 12
   dinosaurs  viii, 92, 93
   mammoths  vii, 3, 4, 5, 6, 11, 12, 13, 21, 28, 29, 74, 81, 95, 96, 103, 105
   sabre-toothed tiger  11, 26
Antarctic  40, 46, 52, 54, 61, 97, 99, 100
asphalt pits  11
asteroids  93, 94, 135, 147, 148
astronomy  161
Atlantic Ocean  161
axis shift of the Earth  xii, 28, 32, 60, 71, 74, 95, 101
   equatorial bulge  22, 23, 24, 33, 35, 62, 65, 72, 96, 98, 101, 139, 141, 142

## B

Babylonian Year  86
Barker, Jason  127
black holes  124, 125, 126, 127, 128, 130, 147
Blytt, Axel  75
Bode, J.E.  135
Bode's Law  135
Brunhes, Bernard  55
Buckland, Professor  8

## C

calendars  viii, 87
   days in the year  88, 91
   Julian  90
carbon dioxide levels  42, 43, 44, 98, 102
Alaska  7, 8, 13, 20, 53, 159
catastrophes  4, 17, 19, 74, 76, 83, 103
   China  10, 17, 18, 81
   Egypt  78, 83, 84, 159
   England  9, 161
   Greenland  5, 45, 47, 51, 52, 53, 82, 87, 99
   Mediterranean  9, 27, 107
   Mexico  8
   Santorini  2, 17
   Siberia  3, 4, 6, 13, 53, 74, 103
cave disasters  11
Ceres  148
climate  viii, 69, 76, 158, 160
   glaciation  9, 15, 43, 48, 51, 52, 53, 76
   global climate  7
   Holocene weather  viii, 74
   Ice Ages  38, 157, 158, 159
   Siberia  3, 4, 6, 13, 53, 74, 103
   temperature changes  41, 45, 48, 94
   warming  7, 64, 98, 99, 100
comets  116, 117, 119, 143, 144, 145, 147, 149, 151
Continental Drift  161
Coriolis effect  74, 75
Cro-Magnon man  106, 107

## D

Darwin, Charles  4, 21
   extinction of mammoths  4
   the Origin of Species  157
Darwin, Sir George  21

de Seres, Marcel  9
desertification  xii, 25
dinosaurs  viii, 92, 93
doomsday  viii, 95, 152

**E**

Earth  vii, viii, xi, xii, xiv, 1, 2, 3, 13, 21, 22, 23, 24, 25, 29, 30, 31, 32, 33, 34, 35, 36, 37, 38, 40, 41, 42, 43, 44, 45, 47, 49, 51, 55, 56, 57, 58, 59, 60, 61, 62, 63, 64, 65, 68, 71, 72, 73, 74, 75, 76, 77, 78, 79, 80, 82, 83, 84, 85, 86, 87, 88, 90, 91, 92, 93, 94, 95, 96, 97, 98, 99, 100, 101, 102, 103, 104, 106, 115, 120, 124, 129, 130, 131, 132, 133, 134, 135, 136, 139, 148, 152, 154, 157, 158, 159, 161
  axis shift  xii, 28, 32, 60, 71, 74, 95, 101
  Coriolis effect  74, 75
earthquakes  vii, 2, 16, 18, 19, 58, 96, 106
equatorial bulge  22, 23, 24, 33, 35, 62, 65, 72, 96, 98, 101, 139, 141, 142
  magnetic field  vii, viii, xii, 1, 23, 33, 36, 37, 55, 56, 57, 58, 59, 60, 61, 62, 63, 64, 65, 68, 97, 98, 99, 100, 101, 104, 116, 121, 122, 129, 155
  movements  xiii, 19, 25, 47, 48, 59, 81, 82, 86, 96, 102, 114, 121, 122, 152
  pole shift  vii, viii, 25, 28, 31, 36, 37, 45, 46, 47, 51, 52, 63, 65, 66, 69, 72, 80, 86, 87, 88, 96, 97, 98, 99, 101, 102, 104, 105, 113
  rotation  xi, xiv, 1, 21, 22, 23, 29, 30, 31, 32, 33, 34, 36, 47, 59, 60, 61, 62, 64, 65, 68, 74, 76, 77, 78, 82, 83, 87, 88, 90, 91, 95, 97, 98, 100, 102, 104, 128, 131, 132, 133, 139, 140, 149, 151, 152, 153
  seabed sinking  19
Earth's rotation  xiv, 1, 23, 31, 32, 33, 34, 36, 47, 60, 64, 65, 74, 77, 98, 104, 132
Eddington, Arthur  34

Egypt  78, 83, 84, 159
Einstein, Albert  35, 158
Elsasser, Walter  59
Emiliani, Cesare  71, 94
Epstein, Samuel  54
equatorial bulge  22, 23, 24, 33, 35, 62, 65, 72, 96, 98, 101, 139, 141, 142
Ericson, David  71
Ewing & Donn, 43

**F**

floods  xi, 77
Folgheraiter, Giuseppe  57

**G**

Gaskell and Morris  41, 47
geographical poles  23, 61, 65, 74, 87, 94, 95, 104
global warming  98, 99, 100
Gold, Dr Thomas  24
gravity  viii, 114, 115, 116, 117, 118, 119, 120, 121, 122, 124, 125, 128, 133, 144, 145, 147, 148, 153, 154
greenhouse gases  42, 43, 98
greenhouse warming  7
Gulf Stream  37, 39, 43

**H**

Halley's Comet  144
Hapgood, Professor Charles  35
Harrison, E.R.  126
Harwood, J.M.  63
Hein, Arnold  15
Herodotus  77, 80, 94, 159
Himalayas  15, 18
Holocene weather  viii, 74
Hoyle, Sir Fred  4

Hutton, James 1

## I

ice ages vii, 26, 34, 37, 39, 41, 42, 43, 46, 48
    and pole shift vii, viii, 25, 28, 31, 36, 37, 45, 46, 47, 51, 52, 63, 65, 66, 69, 72, 80, 86, 87, 88, 96, 97, 98, 99, 101, 102, 104, 105, 113
    glaciation 9, 15, 43, 48, 51, 52, 53, 76
    Laurentide Ice Sheet 46, 48
    Little Ice Age 42, 47, 48
    polar ice vii, 37, 97, 98, 101
    sea levels vii, 48, 51
Imbrie, John and Katherine 38, 54

## J

Jacobs, J.A. 58
Julian Calendar 90
Jupiter 93, 122, 136, 137, 138, 140, 141, 142, 145, 146, 147, 148

## K

Kepler, Johannes 86
Kukla, George 97

## L

Lake Titicaca 14, 111, 112, 113
Lamb, Professor H.H. 76
Laplace theory 156
Larsen, James 127
Laurentide Ice Sheet 46, 48
lava 2, 33, 56, 57, 64, 65
    magnetic alignments 56, 66
    seafloor spreading viii, 44, 64
Lyell, Sir Charles 2

## M

Magnetic Centre of the solar system  128
   alignments  56, 58, 66
   dipolar component  62, 63
   intensity  41, 59, 62, 126
Magnetic  vii, viii, 55, 104, 121, 122, 123, 124, 126, 128, 129, 133, 134, 137, 142, 144, 146, 147, 148, 150, 151, 153, 156, 159
magnetic field of the Earth  36, 56
   recent reversals  67
   seabed reversals  65
Malin, S.C.R.  63
   dwarf mammoths  95, 105
mammoths  vii, 3, 4, 5, 6, 11, 12, 13, 21, 28, 29, 74, 81, 95, 96, 103, 105
   sudden death  vii, 3
Markham, Sir Clements  14
Matuyama, Motonori  56
Maunder Minimum  42, 129, 130
megalithic structures  107
Mercanton  57, 84, 86, 88, 160
Mercury  117, 130, 133, 134, 136, 143
Milankovitch, Milutin  41
Milky Way  122
Moon  139, 142
mountain building  19, 21

## N

NASA  146, 147
Neptune  122, 135, 136, 141, 142
   Sunspots  ix, 129, 158
   law of motion  114, 118
Newton, Sir Isaac  114
   theory of gravity  114, 115
   location  xi, 23, 28, 41, 45, 46, 47, 53, 54, 56, 57, 61, 66,

70, 72, 86, 88, 130, 133, 150
North Pole  13, 28, 30, 34, 37, 45, 53, 56, 57, 61, 62, 64, 81, 83, 86, 87, 96, 99, 131, 136, 138
   in Davis Straits  52

## O

Oort, Jan  143

## P

peat bogs  75
Persian year  89
Pineault, Serge  127
Pioneer 10 and 11  ix, 146, 147
   Bode's Law  135
   Jupiter  93, 122, 136, 137, 138, 140, 141, 142, 145, 146, 147, 148
   locations  ix, 13, 21, 48, 50, 66, 74, 134, 140
   Mars  93, 130, 135, 139, 140, 141, 147, 148
   Mercury  117, 130, 133, 134, 136, 143
   momentum  xiv, 1, 23, 24, 34, 36, 114, 115, 116, 117, 118, 119, 144, 145, 151, 152, 153, 154, 155, 156
   Neptune  122, 135, 136, 141, 142
   orbits  117, 118, 126, 132, 133, 136, 137, 139, 140, 141, 143, 144, 145, 148, 149, 150
   planets  viii, ix, xiii, 23, 87, 114, 115, 116, 117, 118, 122, 123, 124, 127, 132, 133, 134, 135, 136, 137, 138, 139, 140, 141, 142, 143, 145, 146, 149, 150, 151, 152, 153, 154, 155, 156
Pluto  117, 128, 134, 135, 136, 146
   Saturn  122, 135, 141, 142, 146
   Uranus  32, 60, 122, 132, 135, 139, 150
   Venus  30, 60, 132, 150
Plato, pole shift  77, 78, 80, 159, 160
Pogo, A.  84
polar ice  vii, 37, 97, 98, 101

polar excursions  65, 68
geographical poles  23, 61, 65, 74, 87, 94, 95, 104
North Pole  13, 28, 30, 34, 37, 45, 53, 56, 57, 61, 62, 64, 81, 83, 86, 87, 96, 99, 131, 136, 138
poles  vii, 21, 23, 24, 29, 30, 31, 32, 34, 36, 37, 38, 39, 43, 46, 47, 48, 51, 52, 54, 55, 57, 60, 61, 63, 65, 68, 71, 73, 74, 75, 86, 87, 88, 90, 94, 95, 97, 98, 99, 102, 103, 104, 112, 138
pole shift  vii, viii, 25, 28, 31, 36, 37, 45, 46, 47, 51, 52, 63, 65, 66, 69, 72, 80, 86, 87, 88, 96, 97, 98, 99, 101, 102, 104, 105, 113
   and ice ages  38, 157, 158, 159
   sea levels  vii, 48, 51
Pole Star  84, 85
Posnansky, Professor A.  111
pottery, magnetic alignments  10, 27, 57, 65, 67, 92, 112
Pratt, J.H.  34
Prestwich, Professor Joseph  9

# R

radiation balance of the globe  39
radiocarbon dating  7
Runcorn, Professor S.K.  23

# S

Sacsayhuaman  108, 110
Sahara Desert  vii, 25
Saturn  122, 135, 141, 142, 146
Schott, Walter  69, 71
Schumacher-Levy comet  145
seabed  2, 18, 19, 44, 64, 65, 69
sinking  19
   temperature changes  41, 45, 48, 94
Sea levels  vii, 48, 51
Seneca  85, 161

Sernander, Rutger 75
Siberia 3, 4, 6, 13, 53, 74, 103
   extinction of mammoths 4
   black holes 124, 125, 126, 127, 128, 130, 147
Magnetic Centre viii, 122, 123, 124, 126, 128, 129, 133, 134, 137, 142, 144, 146, 147, 148, 150, 151, 153, 156
movements of planets xiii
origin of ix, 155
solar system ix, xiii, xiv, 23, 42, 93, 114, 115, 116, 118, 119, 121, 122, 123, 128, 131, 133, 139, 143, 147, 149, 150, 151, 152, 153, 154, 155
space exploration
   Pioneer 10 and 11 ix, 146, 147
   Ulysses space probe 137
   corona 137, 138
   Magnetic Centre viii, 122, 123, 124, 126, 128, 129, 133, 134, 137, 142, 144, 146, 147, 148, 150, 151, 153, 156
   orbits 117, 118, 126, 132, 133, 136, 137, 139, 140, 141, 143, 144, 145, 148, 149, 150
   polar reversals 84
Sunspots ix, 129, 158

**T**

tar 11
Tarling, D.H. 68
temperature changes 41, 45, 48, 94
Thompson, Sir William 21
Tiahuanaco viii, 14, 110, 111, 112, 113, 157, 160

**U**

Ulysses space probe 137
uniformitarianism 1
Uranus, change of rotation 32, 60, 122, 132, 135, 139, 150

## V

Van Allen, James  116, 124
Venus, change of rotation  30, 60, 132, 150
Volcanoes  vii, 2, 16

## W

Wallace, A.R.  15
Wegener, Alfred  52, 64
Whipple, Professor  117
Wilson, Alex T.  42
winds  12, 13, 39, 49, 75, 76, 107, 138
   and destruction  xii, 8, 13, 21, 78, 80, 92, 103, 105, 106
   directions  18, 22, 23, 24, 30, 31, 32, 33, 34, 40, 48, 52, 56, 59, 60, 61, 63, 64, 66, 68, 74, 75, 76, 82, 84, 87, 98, 102, 104, 118, 126, 131, 132, 136, 141, 144, 145, 147, 149, 151, 153, 155, 157, 160
   solar wind  23, 25, 30, 31, 32, 33, 34, 36, 37, 60, 63, 74, 82, 87, 91, 95, 97, 98, 99, 100, 101, 102, 116, 123, 124, 128, 131, 132, 133, 134, 135, 136, 137, 138, 139, 144, 145, 147, 148, 149, 154, 156
wind movement  75, 106
Wrangel Island  28, 95, 96, 105

## X

X-rays  126, 130

## Y

Yapp, Crayton  54
Yellowstone National Park  17

## Z

Zeilik, Professor Michael  125

www.ingramcontent.com/pod-product-compliance
Lightning Source LLC
Chambersburg PA
CBHW052030070526
44584CB00016B/1974